SpringerBriefs in Applied Sciences and Technology

SpringerBriefs present concise summaries of cutting-edge research and practical applications across a wide spectrum of fields. Featuring compact volumes of 50 to 125 pages, the series covers a range of content from professional to academic.

Typical publications can be:

A timely report of state-of-the art methods
An introduction to or a manual for the application of mathematical or computer techniques
A bridge between new research results, as published in journal articles
A snapshot of a hot or emerging topic
An in-depth case study
A presentation of core concepts that students must understand in order to make independent contributions

SpringerBriefs are characterized by fast, global electronic dissemination, standard publishing contracts, standardized manuscript preparation and formatting guidelines, and expedited production schedules.

On the one hand, SpringerBriefs in Applied Sciences and Technology are devoted to the publication of fundamentals and applications within the different classical engineering disciplines as well as in interdisciplinary fields that recently emerged between these areas. On the other hand, as the boundary separating fundamental research and applied technology is more and more dissolving, this series is particularly open to trans-disciplinary topics between fundamental science and engineering.

Indexed by EI-Compendex and Springerlink

More information about this series at http://www.springer.com/series/8884

Mrinal Kaushik
Prashanth Reddy Hanmaiahgari

Essentials of Aircraft Armaments

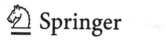 Springer

Mrinal Kaushik
Department of Aerospace Engineering
Indian Institute of Technology Kharagpur
Kharagpur, West Bengal
India

Prashanth Reddy Hanmaiahgari
Department of Civil Engineering
Indian Institute of Technology Kharagpur
Kharagpur, West Bengal
India

ISSN 2191-530X   ISSN 2191-5318   (electronic)
SpringerBriefs in Applied Sciences and Technology
ISBN 978-981-10-2376-7   ISBN 978-981-10-2377-4   (eBook)
DOI 10.1007/978-981-10-2377-4

Library of Congress Control Number: 2016948758

Printed on acid-free paper

This Springer imprint is published by Springer Nature
The registered company is Springer Science+Business Media Singapore Pte Ltd.

To
My sweet son 'Talin' and
my little angel 'Tanvi'
and My wife 'Sangeeta'

—Mrinal Kaushik

Preface

With great pleasure, this book on aircraft armaments is delivered. It is written for all those who are either interested in studying the subject as beginner or working in the armed forces and interested to use it as reference. The book is a compilation of lectures delivered by me to the armed forces officers stationed at *Defence Institute of Advanced Technology*, Pune and the post-induction trainee engineers of *Hindustan Aeronautics Limited*. Dr. Prashanth Reddy Hanmaiahgari joins Dr. Mrinal Kaushik as co-author in this edition.

The book begins with introducing elementary concepts about various armaments launched by the combat aircrafts in Chap. 1. At the end some references are also provided so that interested readers may get more details about these 'air-dropped-ammunitions.'

In Chap. 2, the fixed and flexible guns mounted on aircraft are discussed. To intercept combat aircrafts, a brief review on anti-aircraft guns is also presented.

The detailed chronological development of bombs is discussed at a length in Chap. 3. It begins with the discussion on the '*general-purpose bombs*' and subsequently provides a detailed treatment on '*atomic*' and '*thermo-nuclear*' bombs.

In Chap. 4, a detailed and consolidated account of cutting-edge technology of missiles is presented. Broad classification of missiles depending upon their *Type, Launch Mode, Range, Propulsion System, Warhead Used,* and *Guidance System* are written in such a manner that even an average student can grasp them without facing any difficulty. Subsequently the state-of-art treatment on '*ballistic missiles*' is also attempted. The chapter ends with highlighting problems associated in missile launch.

The brief discussion on armament materials is given in Chap. 5.

The organization of United Nations (UN) is briefly discussed in Chap. 6. Prevention of nuclear, biological, and chemical weapons through various treaties and conventions as adopted by the UN General Assembly are highlighted at the end.

A large number of multiple choice questions with their answers are given in *Appendix* for the readers to evaluate their understanding.

I am confident that the readers will find the manuscript useful and will be able to grasp the fundamentals easily.

Kharagpur, India   Mrinal Kaushik
June 2016

Acknowledgements

First of all, I thank the almighty for providing me the courage which brought this book into reality. My life is extremely indebted to Dr. Prahlada, Padma Shri awardee by Government of India in 2015, and former vice chancellor of Defence Institute of Advanced Technology (D.I.A.T.), Pune, for appointing me as Deputy Program Director of Post Induction Training School (POINTS) of Defence Research and Development Organization (D.R.D.O.), Pune. While working at this position, I had an opportunity to interact India's leading missile scientists and armed forces officers which helped me in gaining the knowledge about the armaments. I am really thankful to Dr. S.E. Talole, Dean (Technology) and Dr. Ajay Misra, Head Aerospace Engineering Department at D.I.A.T., Pune for their facilitation in teaching missile aerodynamics course during my stay with them.

Another special personality which must be acknowledged here is none other than my best friend Dr. Rakesh Mathpal, Assistant Professor at Department of Aerospace Engineering of Indian Institute of Technology, Kanpur and former research scientist of Indian Space Research Organization (ISRO), Trivandrum for his generous support during all odd and even situations in my life. He helped me in grasping the technology behind the geosynchronous- and polar satellite launch vehicles (GSLV and PSLV), development while working with him at ISRO, Trivandrum.

I wish to thank my co-author Dr. Prashanth Reddy Hanmaiahgari, who supported me by critically checking the manuscript and making it error-free to a large extent; my doctoral student Humrutha for giving useful suggestions during the preparation of the manuscript.

The financial support provided by Continuing Education Programme (CEP) of Indian Institute of Technology, Kharagpur in writing the text is also deeply acknowledged.

I will fail in my duty if I do not especially acknowledge to all those friends who studied with me during my Ph.D., M.Tech, and B.Tech degrees at Indian Institute of Technology, Kanpur. Their loving memories are an asset in my life.

Last but not least, it is my pleasant duty to acknowledge the help provided by Ms. Swati Meherishi and Ms. Aparajita Singh of Springer in preparing the manuscript.

Kharagpur, India   Mrinal Kaushik
June 2016

Contents

About the Authors

Dr. Mrinal Kaushik is Assistant Professor at the Department of Aerospace Engineering at IIT Kharagpur, India. He has earned his Ph.D., M.Tech, and B.Tech degrees all in Aerospace Engineering from IIT Kanpur, India. His primary area of expertise is experimental supersonic aerodynamics with the current focus on Active and Passive Control of Shock-Boundary Layer Interactions and Active and Passive Control of Jets and Base Flows. His other research includes experimental investigations on Aerothermodynamics of Hypersonic Vehicles and Hydrodynamics of Hydrofoils. For his contributions in technology, he has received 'Young Researcher Award' from Venus International Foundation (India) in 2015. Dr. Kaushik's biography is included in the prestigious database 'Marquis Who's Who World' in 2016.

Dr. Prashanth Reddy Hanmaiahgari is Assistant Professor at the Department of Civil Engineering in IIT Kharagpur, India. He has obtained his Ph.D. and M.Tech degrees in Civil Engineering from IIT Madras, India. Dr. Reddy has also worked as postdoctoral fellow at the reputed universities in USA and Canada for several years. He works in the domain of computational and experimental hydrodynamics. His research is multidisciplinary and involves fluid mechanics across mechanical and civil engineering. Dr. Hanmaiahgari's biography is included in the prestigious database 'Marquis Who's Who World' in 2010.

Chapter 1
Basic Concepts

Abstract In this chapter, the weapons used in the act of warfare are introduced. However, the discussion is confined only to the munitions which are launched from a combat aircraft such as aircraft guns, air-dropped-bombs, guided missiles, etc. The broad classifications of missiles are also briefly introduced.

Keywords Weapon · Armament · Munitions · Aircraft guns · Air-dropped bombs · Missiles · Combat aircraft

Weapon is any means by which one contends against another. Armament is a mechanism that enables war fighters to inflict damage to enemy targets to a degree that will inhibit the enemy's ability to engage in the further act of warfare. All the equipment through which a combat aircraft can release destructive power on a target can be termed as air Armaments or Ordnance. Armament is generally termed as the process of equipping and supplying war weapons to the military. These are categorized based on their function and the way they are used or aimed onto the target. They are typically delivered by static winged aircrafts or Rotorcrafts. However, nuclear weapons are not included in this category as they derive their explosion force either from nuclear fusion or fission reactions. Weapons with dispensers such as guns and projectiles, landmines, missiles, warheads, naval mines, rockets, free-fall bombs, and cluster munitions are specifically included in this category. Submunitions, air-to-air and air-to-surface guided weapons, mines and anti-radiation missiles, etc., are also included in this group. Nuclear weapons such as air-launched cruise missiles and nuclear bombs delivered by aircrafts, which are considered as strategic weapons, fall under a specific class of air armament. Free rockets or unguided rockets developed during World War II are more accurate than bombs and less accurate than guns. Even though rocket systems are often considered obsolete, virtually all major countries maintain one or more of them in their arsenal. Although not normally included as a part of the field of aircraft armament, fire control is the term that covers the sighting, aiming, and computation which enables the pilot or aircrew to hit the target. Basically, it examines the conditions of the engagement and indicates when to release the

© The Author(s) 2017
M. Kaushik and P.R. Hanmaiahgari, *Essentials of Aircraft Armaments*,
SpringerBriefs in Applied Sciences and Technology,
DOI 10.1007/978-981-10-2377-4_1

armament in order to obtain its goal. The trend in the warfare decides the trends in weapon and armament developments. The concept of warfare has been revolutionized by the modern weapons system and the military technology. The missile guidance technology helps us to understand the various methods of delivering a missile or bomb toward its target accurately which in turn determines its effectiveness. The technology of guided missiles acts as a force multiplier and provides a competitive and cutting edge in improving the missile accuracy depending on the type of target, which in turn is a critical factor in any combat or military operations. It compasses the multiple streams of engineering, technology, and applied sciences. The successful, accurate, and an effective launch of a missile depend on a number of factors involving a coordination of variety of subsystems.

This book is an attempt to provide a feeling of scientific knowledge, awareness, and familiarity with aircraft armaments to the general public.

1.1 Guns [1]

In World War I, the belligerents had a few wood and canvas aircrafts and were initially intended for using as scouts and reconnaissance. The pilots were provided with pistols, rifles etc. to defend themselves in case they spot and locate the enemy troops and to take potshots on the rival aircrafts by dropping grenades on the enemy troops. The air warfare took a major turn with the usage of machine guns fitted on planes with a variety of mountings. Initially, the two-seated aircrafts carrying guns fitted in their rear cockpits were used throughout the war and these machine guns fitted on the planes were used to supply defensive fire. Fire forwarding machine was devised due to the necessity of offensive fire. Pusher planes like The *Airco-DH2* had its engine in the back and the gun mounted in the front. Machine guns are mounted in different angles like some of the planes are equipped with guns on the side wings, upper wing, above and below the cockpit. The machine guns are mounted behind the propeller using *"Fokker's Synchronizer"* (also known as Interrupter) by the latest fighter aircrafts for a better accuracy and this idea helps the aircraft in better and fixed aiming of the armament toward the target instead of aiming it independently.

In the World War II more powerful, portable and lighter machine and submachine guns were manufactured due to the availability of better materials. These guns fire through the propeller spinner and are generally mounted in the wings, nose, and engine cowlings or between the banks of engines of the aircrafts. One to fourteen flexible machine guns and/or auto canon, which have the capability of firing upwards were used as defensive armament by the bombers and night fighters. Fixed offensive guns were also used at times. Since the early 1960s, missiles were used as the primary weaponry but the guns still played a vital role in the Vietnam War. Cannons (typically between 20 and 30 mm in caliber) were fitted on the fighters, built since then and were used as an adjunct to missiles. Gatling guns are favored by the United States and to some extent Russia, whereas the modern fighter aircrafts equipped with the revolver canon are used by the Europeans.

1.2 Bombs [2]

A bomb is a destructive weapon and usually a conglomerate of various explosive materials which has the capability of sudden explosion with a massive release of energy based on exothermic reactions.

Aircraft bombs are usually based on the principle of blasting or fragmentation with an immense release of energy from the point of detonation toward the enemy targets to reduce and neutralize the enemy's war potential. This is done by destructive explosion, fire, war gases and nuclear reactions. Aircraft bomb ammunition is used strategically to destroy installations, combatants, military armaments and personnel, tactically in direct support of land, sea, and air forces engaged in offensive or defensive operations. They usually follow a trajectory based on strategy and tactics. Bombs are designed to be carried either in the bomb bay of aircraft or externally under the wings or fuselage.

The bombs are broadly classified into two categories: general purpose bombs and nuclear bombs. Common examples of general purpose guided weapons are *PAVWAY* series (the USA), *GBU* and *JDAM* series (the USA), *KAB* series (USSR), *LIZARD* series (ISRAEL), etc. The nuclear bombs either work on the principle of 'nuclear fission' or 'nuclear fusion,' The first category includes atomic bombs and famous examples of this type are '*Little Boy*' dropped on Hiroshima and '*Fat Man*' dropped over Nagasaki during the Second World War. The fusion bombs are also called as '*thermonuclear bombs*' or '*Hydrogen bombs*' and are much more powerful compared to their fission counterparts.

1.3 Missiles [3]

Missile is a smart war weapon or any ammunition which is designed to hit a specific target with low collateral damage. It is usually designed with an automatic guidance system that facilitates it to identify and track its own current position, the path it is supposed to take, the position and mobility of the intended targets, the distance between them and ultimately self-designs the course of is trajectory. Their path is guided and controlled from the launch to the terminal stage with or without propulsion. Missiles do have an engine and are equipped with warheads thus providing them with an ability of inflicting lethal destruction power/damage to the intended target. Missiles differ from the rocket by virtue of a guidance system that steers them toward a preselected target. They have a flight system which enables them to collect data from the guidance system. They are equipped with explosive warheads (destructives materials) and other various forms of chemical, radiological, and biological agents and are used in warfare to destruct the intended target. The combination of kinetic energy and high speeds also contributes to their destructive power and allows them to hit some tough targets.

Guided missiles are mainly operated on the technologies of Propulsion (the required energy source provided for its movement), Guidance (capacity and intelligence to go in the right path and direction), and Control (effective maneuvering). These factors help in calculating the size, range, and state of motion of missiles and make them more targets specific.

A broad classification of missiles is based on certain features like the type of engine or rocket used, the nature of the intended target, the type of their launching platform, the method of their flight system, propulsion or guidance, and the type of warheads used. They are further subclassified based on their range and the exact type and nature of their target. Strategic Missiles, Tactical Missiles, Cruise Missiles, and Ballistic Missiles are the common and popular types of missiles used. However, the most usual is the one in which the position of launch and the position of the target are used for classification. This is most widely used as these positions more or less designates the general requirements or specialties of the missiles.

1.4 Summary

It is evident that modern warfare is based on the concepts and methods of advanced military expertise, especially after World War II. These concepts have become more complex and innovative due to the widespread application of advanced information technology in their manufacturing and designing. The war policies have changed gradually as the combatants adopted more modernized concepts of warfare based on strategies, tactics, and operations to preserve their battle competency. Rail guns, massive ordnance air burst bombs, stinger missiles, e-bombs, cluster bombs and rocket-propelled grenades, nuclear bombs; cruise missiles, patriot missiles etc. are the examples of modern warfare activities.

A rail gun is kind of projectile which neither uses explosives nor propellants but uses electromagnetic forces to attain a high speed with which it can hit a target of 250 miles away within a few seconds. Rocket propelled grenades are effective against helicopters, armored vehicles, attacking buildings at a closer range, and disabling war tanks. Nuclear weapons are powerful and result in a mass destruction and cause a severe threat and after effects to the mankind, hence used in various conflicts. Embedded or hidden underground targets are targeted by weapons like Bunker busters, which have a deep penetrating capacity unlike ordinary bombs. Cruise missiles have good accuracy and are conceptualized on remote battle technology, often used in modern warfare. Missile is essentially a powerful unmanned airplane which can deliver the bombs weighing up to tons to the targets located thousand miles away. Smart bombs have startling accuracy and can hit specific ground targets even in bad weather conditions. They are examples of the leading technology in recent days. E-bombs are electromagnetic weapons that use electromagnetic field to destroy the electrical devices of the enemy which contain any important data. The newest, largest, and the most deadly conventional bomb ever built by the US arsenal is the Massive Ordnance Air Burst bomb which is considered to be the *"mother of all bombs."*

References

1. Bruce JM (1965) Warplanes of the First World War—fighters, vol 1. MacDonald & Co., London
2. Needham J (1986) Military technology: the Gunpowder Epic. Cambridge University Press, Cambridge, pp 180–181
3. Nielsen JN (1960) Missile aerodynamics. Nielsen Engineering & Research, Inc., California

Chapter 2
Guns

Abstract In China around AD 1000, the tubular bamboo stick was used to launch spikes by burning gun-powder. The combustion generates compressed gas at high pressure which was used in ejecting spikes at high speed. The gun powder is a mixture fuel and oxidizer. Generally, the sulphur and charcoal are used as fuel and potassium nitrate as oxidizer. This chapter briefly reviews the use of fixed and flexible guns mounted on the aircraft.

Keywords Aircraft gun · Anti-aircraft gun · Gatling gun · Cannon calibre · Fixed gun · Flexible gun

A gun is an air-to-air, tubular weapon used to discharge projectiles and other materials with high velocity. It is by far the most widely used weapon ever devised. Aircraft guns are generally classified as either *Fixed* or *Flexible*.

Fixed guns are installed in a stationary position and are not movable in other directions unrelated to the aircraft. They are usually forward firing and the entire aircraft (fighter) must change its direction to move the weapon and aim it. Fixed, forward firing guns usually need a single operator to aim and steer and have lighter installations, produce less drag and hence have less negative impact on their performance. This makes them most the favorable weapons for small, maneuverable fighters [1].

Flexible guns on the other hand are fixed to a platform on the aircraft but can be rotatable to cover a certain field of fire and can be aimed up and down, side to side directions and at certain elevations by the operator irrelative to the direction of the vehicle. Such guns are installed and manually operated in power turrets. In addition to the pilot, flexible guns require another dedicated operator for aiming the weapon, thus adding to the size and weight of the aircraft. When the opponent can be kept in front of the attacker, maneuvering relative to another aircraft which essentially requires a forward field of fire becomes much simpler. For these reasons, flexible guns are generally preferred for the defense of larger, less maneuverable aircrafts whereas the fixed forward-firing guns have been found to be more advantageous for small, offensive aircrafts (fighters).

© The Author(s) 2017
M. Kaushik and P.R. Hanmaiahgari, *Essentials of Aircraft Armaments*,
SpringerBriefs in Applied Sciences and Technology,
DOI 10.1007/978-981-10-2377-4_2

In World War I, the fighter armaments such as personal side arms and weaponry were improvised and progressed to flexible machine guns and eventually to fixed machine guns using innovative technology. The standard fighters were usually equipped with synchronizers to allow fire through the propeller disc. But by the end of this conflict, they were improvised to two 0.30 calibre class fixed forward-firing machine guns.

A Cannon is essentially a kind of gun in the form of tube which uses explosive materials to launch a projectile. It varies in mobility, range, angle of firing, size etc. The development of aircraft cannons was based on the search for more destructive projectiles. Generally these explosive charges on contacting the target explode as they are armed by the firing acceleration of the shell. In World War I, single-shot cannons were used to some extent but the true, effective automatic cannons were developed between the wars. These cannons were generally 20–40 mm weapons and have greater destructive power than the machine guns. They were larger and heavier in size which lead to the further tradeoffs in usable aircraft space and in performance. With the correspondingly lower rates of fire, these cannons had projectiles significantly larger than those of the 0.30 and 0.50 calibre class, commonly used machine guns.

After World War II, a new significant technological breakthrough has appeared in air-to-air guns. A cannon with new design known as the 'M39' was built in the United States. It's built around a rotating cylinder similar to a "revolver" handgun and modeled from an experimental German gun which resulted in a greater increase in the rate of fire (Fig. 2.1).

'Gatling gun' cannons were introduced in the later 1950's with greater performance capabilities [2]. It was designated as 'M61' in the United States. This gun could develop a tremendous rate of fire with less barrel overheating and erosion and it was employed with a multiple rotating barrels rather than a revolving cylinder.

Fig. 2.1 Gatling gun used in American war

In addition to it, this gun was usually propelled electrically, hydraulically or pneumatically. Also problems associated with duds were eliminated as it was not dependent on the residual energy of the expended round.

The trend of the gun being the fighter's primary armament saw a definite change in the 1950's and the 1960's. Many of the fighters were not equipped with guns at all during this period. It's because of the feeling that the high speeds of jet fighters and the heavy armament of new bombers, particularly suitable for the night and all-weather missions has made the gun obsolete. The package of air-to-air weapons consisted entirely of guided missiles and unguided rockets. But later on, the importance and usage of guns has become prevalent once again due to the limitations of some of the more exotic weapons. Further combat experiences had once again demonstrated the value of the gun thus reversing the trend in the 1970's.

References

1. Bruce JM (1965) Warplanes of the First World War—fighters, vol 1. MacDonald & Co., London
2. Needham J (1986) Military technology: the Gunpowder Epic. Cambridge University Press, Cambridge, pp 180–181

Chapter 3
Bombs

Abstract To destroy the enemy's warfare ability, an explosive weapon also called a 'bomb' has been used since time immemorial in the history of human civilization. 'Air-dropped-bombs' are those weapons which are dropped in air targeting the enemy state. The 'grenades' or 'moderate-destructive-explosives' were the first in this category which were dropped by the 'heavier-than-air' aircrafts. Later on, more powerful bombs such as 'atomic,' 'hydrogen,' and 'neutron' bombs were developed in order to have mass destruction.

Keywords Detonation · Aerodynamic decelerators · Cluster bomb · Atomic bomb · Thermonuclear bomb · Neutron bomb

Based on their application, the bombs can be classified in many ways such as fire weapons, war gases, destructive explosions of moderate level, and the disastrous nuclear weapons. The first three are summed up into 'general purpose bombs,' whereas the nuclear weapon category includes 'atomic bomb,' 'hydrogen (H-bomb) or thermo-nuclear bomb,' and 'neutron bomb.'

3.1 General Purpose Bombs

A bomb is basically an explosive and destructive weapon. The extreme, sudden, and violent release of energy from a bomb is due to the result of exothermic reactions of the explosive materials in it. When a bomb detonates, the damage is principally inflicted through the impact and penetration of pressure driven projectiles, ground and atmosphere transmitted mechanical stress, damage due to pressure, fragmentation and shattering of projectiles, and explosion generated effects.

An 'aerial' bomb is an explosive weapon, usually designed to be dropped from an aircraft. It travels through the air and has predictable trajectory and produces an immense destruction to the target by an outward fragmentation from the point of detonation, when hit. Aerial bombs include wide range and complexity of designs

© The Author(s) 2017
M. Kaushik and P.R. Hanmaiahgari, *Essentials of Aircraft Armaments*,
SpringerBriefs in Applied Sciences and Technology,
DOI 10.1007/978-981-10-2377-4_3

based on their mechanism such as '*Glide*' bombs (which take a path), '*Gravity*' bombs (free-fall or dumb bombs), '*Delay-action*' bombs, etc. The concept of aerial bombing includes a wide variety such as unguided gravity bombs to guided bombs, hand tossed from a vehicle to needing a large specially built delivery vehicle or perhaps be the vehicle itself such as a glide bomb and instant detonation or delay-action bombs. Aerial bombs are designed to destroy material, kill and injure people through the blast projection and outward fragmentation.

A bomb is typically designed with a device known as '*Fuze,*' which initiates or triggers the detonation. A bomb fuze is a mechanical or electrical device and consists of elements that signal the detonation. It also has some sensitive explosive elements (the primer and detonator) that initiate the detonation at proper time, after certain conditions are met. The fuze is essentially designed with safety and arming mechanisms and protects the users from accidental or premature detonation. The primer is fired by the mechanical action or an electrical impulse, which causes the detonator to explode and a Booster relays the primer detonator explosion to the main charge.

In 1849, the Austrians launched their first bombs against Venice by unmanned balloons carrying a single bomb. These bombs are delivered to their targets in air. A 'heavier-than-air' aircrafts were used to drop the very first bomb in the form of 'grenades' or 'grenade like devices.' During the Italo-Turkish War, Giulio Gavotte historically made the first use of air drop bombs on November 1, 1911. On the suggestion of the Bulgarian Air Force pilot, *Christo Toprakchiev*, aircrafts were used to air drop bombs (called grenades in the Bulgarian army at that time) on Turkish positions in 1912, during the First Balkan War. The idea of creating several prototypes by adapting different types of grenades and increasing their payload was developed by Captain *Simeon Petrov*. A final design, with an improved aerodynamics, an X-shaped tail and an impact detonator was created by Captain Petrov after number of tests. Until the end of World War I, Bulgarian Air Force widely used this version during the siege of Edirne and remained in mass production till the end of the war. These bombs created a crater of 4–5 m wide and about 1 m deep on impact and weighed about 6 kg. For the first time in this campaign, two of these bombs were dropped on a Turkish railway station from an Albatross F-2 aircraft, by an observer *Prodan Tarakchiev* on October 16, 1912.

An '*unguided bomb*' is usually delivered by conventional aircrafts and does not have any guidance system and hence follows a ballistic trajectory. It is briefly guided during the initial release by the flight weapon system, but its trajectory is subjected to changes to the external forces such as pressure of gases in the atmosphere and gravitational forces (Laws of classical mechanics). It is also known as a free-fall bomb, gravity bomb, dumb bomb, or an iron bomb. The conventional bombs employing TNT or TNT-based mixtures have been made in many types and sizes. The casings for unguided bombs are typically aerodynamic in shape, often with fins at the tail section which reduce drag and increase stability after release, both of which serve to improve accuracy and consistency of trajectory. Unguided bombs on impact would detonate with the help of a contact fuse. In some cases, the penetration into the target surface would trigger the detonation after a few milliseconds.

The other class of bombs termed as '*Retarded-Bomb*' uses mechanical methods in increasing aerodynamic drag. Aerodynamic decelerators are devices which are designed to produce large drag. The decelerators can also be used in dropping missiles over the enemy territory. The drag on an entry vehicle can be increased to a great extent when equipped with inflatable devices called '*Inflatable Aerodynamic Decelerators* (IAD).' These devices are inflated by an internal gas generating source or ram air (or both) and the shape is maintained by a mostly closed gas-pressurized three-dimensional body. When exposed to the high dynamic pressure free stream, the IAD outfitted with inlets inflates in the method of the ram air inflation. Generally, IADs inflate without issue among the atmospheric inflation methods, and the trailing IADs have all been ram air inflated. Tanks of pressurized gas and internal gas generators are required for inflation via an internal gas source. Due to the lack of dynamic pressure and due to the high torus pressure requirements, the exo-atmospherically inflated IADs and the cone IADs need the scheme called 'the internal gas source inflation' (Fig. 3.1).

The *armor-piercing bombs* are a kind of aerial bombs designed for the use against concrete fortifications and submarine pens (Fig. 3.2). Instead of explosives, these bombs are made of dense and inert material, typically concrete, and are usually strong and thick with pointed tip. The momentum and the kinetic energy of the falling bomb help in the destruction of the target on contact. The pointed tip of these bomb helps in penetrating into the hardened targets such as warships, tanks, and bunkers. These bombs should be either configured as laser-guided or should be used in the form of a smartbomb to be practically deployed due to the chances of significant damage when it is hit directly on to a small target. To reduce the civilian casualties and related damages, these bombs are used typically in urban areas to destroy the military vehicles and artillery pieces.

A *demolition bomb*, also known as a *light case bomb,* is another type of general purpose bomb specifically designed for demolition work. The blast effect is the primary source of explosive energy which helps these kinds of bombs to accomplish their mission. The bomb's weight contributes to 75 % of the detonating charge in some of the very large light-cased bombs. These bombs have thin light metal casing

Fig. 3.1 Aerodynamic
decelerators in bombs

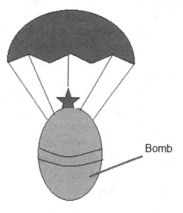

Bomb

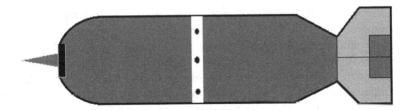

Fig. 3.2 Armour-piercing bomb

and are designed to destruct the target entirely with the energy released during the blast. When the bomb explodes, instant gases with high velocity and temperature are generated. In this mechanism, the metal case of the bomb expands and breaks into pieces (fragmentation) due to the hot gases ejected at the time of detonation and in turn compresses the surrounding air around the case. This compressed body of air is thrown against the adjacent layers of air, which results in the formation of a '*belt.*' The air has high outward velocity and high pressure in this belt. Due to the pressure disturbance, an extremely sharp front called the '*Shock Front*' is formed and it acts as a boundary to this belt. The shock front travels at a high initial velocity from the point of detonation with a large pressure jump across the front. As the shock front travels forward and the pressure decreases, the velocity of the front also decreases rapidly toward the velocity of the sound. Within a short time, the excess pressure prevailing at a point in the air as a result of the arrival of the sharp front gradually decreases and totally vanishes. These are followed by minor disturbances which include partial vacuums. The series of incidents, which produce this entire disturbance in the air at the time of bomb detonation, is termed as a '*blast.*' The *M118* is an air dropped demolition bomb used by US military forces.

A **General-purpose (GP) bomb** is an air-dropped bomb intended to travel in air with a trajectory. They have vast range and a wide design complexity and usually are a compromise between armor-piercing and demolition bombs. In an explosive effect, the general purpose bombs stand as a compromise between fragmentation, blast destruction, and penetration. The common example is incendiary bombs. Incendiary weapons or bombs are purposefully designed to cause maximum fire damage to the target by initiating and spreading fire. They are used to destroy sensitive equipment, structures, and burn supplies. In this case, the targets should be susceptible to fire damage and should partially or completely be made of flammable material so that the fire can spread. Thermites (TH), Magnesium (MG), and combustible hydrocarbons, such as oils and thickened gasoline, chlorine tri-fluoride, white phosphorus, etc., are the main incendiary agents used. Incendiary bombs, in fact, are not explosives, though colloquially called bombs and basically work on the principle of ignition rather than detonation. Instead of detonating instantly, they slow down the process of chemical reactions and produce intense, localized heat to ignite the adjacent combustible target materials. Thus, the target is either destroyed by the initial action or by spreading and continuation of the fire. For example, Napalm, a petroleum product, which is thickened with certain chemicals into a gel kind of product. In this case, the gel adheres to the surface and resists suppression. It does not stop the combustion but slows down the reactions thus releasing energy over a larger time frame unlike other explosive devices. The *Mk 80* series bombs (USA) are typical examples of modern general purpose bombs.

Cluster Bombs: An entirely new class of aircraft bombs known as cluster bombs has evolved in early 1950s (Fig. 3.3). It is partly an outgrowth of cluster bombs from World War II and partly a result of the realization that many targets, notably personnel, and light vehicles could be destroyed more efficiently with several small bombs than with one large bomb. The primary advantage of cluster bombs over unitary bomb is their large footprint or area covered, which compensates for delivery errors and target uncertainty errors incurred by unguided weapons.

A cluster bomb or a cluster munitions is basically a hollow canister consisting of a dozen to few hundreds of submunitions within it. They are designed to be either launched from ground (military tanks combat vehicles), sea (submarines and warships), or air (helicopters and fighter planes). The containers open in the midair and release all the bomblets, and sometimes they are provided with dispensers and are retained by the aircraft once they release the bomblets. The scattered bomblets explode when they hit the ground or the target. Cluster bombs are generally used to attack multiple targets (like vehicles, army troops, infrastructure, etc.) and can cover wider areas of the target sometimes up to several hectares. Once released, the small parachutes fitted to the submunitions retard their speed and allow the aircraft to escape the blast site, in case of low altitude attacks.

The advancement in the bombardment strategy together with its accuracy had been the result of difficult detection, and the discoverable airplanes and mainly the accurately guided bombs being determined against the ground targets. Therefore, the bombing accuracy together with the plane safety is very important characteristic of contemporary air forces. The first precise guided bombs were used by US Air Force in the Vietnam War, known as the bombs of *PAVEWAY-I* series. The laser-guided bombs thrown away from F–14 planes had destroyed the *Thanh Hoa Bridge*. The amount of these bomb types (approximately 25,000 pieces) can be considered rather as improvised. These weapons of the first generation were thrown

Fig. 3.3 Photographic view of a typical cluster bomb

from the height of 3000 m during daylight. During the Gulf War, the precise guided systems with laser and television as well as infrared guidance systems were at disposal. Approximately half of them were the precise guided bombs of *PAVEWAY-II* and *PAVEWAY-III* series.

The *KAB* series bombs are a TV-guided, fire and forget, electro-optical type bombs developed by the USSR in 1980s. The *KAB-500KR* is equivalent to the Americans *GBU-15* weapon. The KAB-500KR is 3.05 m in length and out of its total weight of 520 kg, 380 kg wt comprises of armor-piercing, hardened warhead capable of penetrating a reinforced concrete to the depth of about 1.5 m. The weapons seeker can lock onto the target up to 17 km depending upon the visibility.

The Israeli *LIZARD* series laser-guided bombs are suitable for integration with several warhead sizes, the current *LIZARD-IV* laser designator is available in three operating modes, respectively, for against stationary or slow-moving ground targets, faster moving targets or for pinpoint strikes using a combination of INS/GPS guidance and terminal laser inputs.

3.2 Atomic Bomb [1]

The atomic weapons derive their massive destructive power from the principle of either fission or fusion. Certain fissile elements, which make up the bombs core, release immense amount of energy while splitting their nuclei into smaller fragments. During Second World War, two types of atomic bombs are developed by the United States. The first one is designed with a Uranium core and is a gun-type of weapon named '*Little Boy*,' dropped on the Japanese city of Hiroshima. The second one is with a Plutonium core, kind of an implosion-type device named '*Fat Boy*,' dropped on Nagasaki.

Later on, many nations like the USSR (1949; now Russia), Great Britain (1952), France (1960), and China (1964) subsequently have developed atomic bombs. Presently, countries like India, Israel, Pakistan, and North Korea do have the atomic bombs and the capability to develop them. Nuclear arsenals and warheads inherited by the three smaller Soviet successor states (Ukraine, Kazakhstan, and Belarus) were relinquished and moved back to Russia.

3.2.1 Nuclear Fission

Atomic scientists have mainly selected radioactive isotopes like Uranium-235 and Plutonium-239 due to their property of readily undergoing fission. Fission is a process in which a large nucleus is divided into two or more fragments with roughly equal mass of the mother nucleus. Because the mass of the fragments is less than that of the main nucleus, this reduction in the mass comes out as energy with the release of some neutrons. Every fission reaction results in the formation of new smaller nuclei and releases of some neutrons which in turn bombard with the newly formed nuclei and thus continue the fission making it a chain reaction. When a slow moving neutron hits a massive nucleus like Uranium-235, it breaks up into two new atoms with the release of some binding energy and three neutrons. Fission is further continued by only one of these neutrons as remaining two are either lost or absorbed by an atom of Uranium-238. A single neutron collides with an atom of Uranium-235 and continues fission producing smaller nuclei, two neutrons and some binding energy. Again, both these two neutrons released, bombard with U-235 atoms resulting in fission, thereby releasing one or three neutrons each. This process continue as a nuclear chain reaction and finally an atomic explosion

$$^{235}_{92}U + ^{1}_{0}n^- \rightarrow ^{141}_{56}Ba + ^{92}_{36}Kr + 3^{1}_{0}n$$

3.2.2 Energy Release in the Fission

The total amount of energy released in nuclear fission may be calculated from the difference in rest masses of reactants $\left(^{235}_{92}U + ^{1}_{0}n\right)$ and final products $\left(^{141}_{56}Ba + ^{92}_{36}Kr + 3^{1}_{0}n\right)$. The energy released can be obtained by using Einstein mass-energy relation $E = MC^2$, where E is the released energy, M is mass of fissile material and C is speed of light (Table 3.1).

3.2.3 Criticality

A critical mass is defined as the minimum amount of fissile material required for sustaining a chain reaction in order to detonate an atomic weapon. The critical mass of the material is influenced by various factors like density, temperature, shape, and surroundings. In a chain reaction, to ensure that neutrons released by fission will strike another nucleus, enough U-235 or P-239 is needed. On an average, the amount of the material (fissile) required to initiate a fission reaction and produce neutrons that have the capability of continuing it, instead of going waste, absorbed or leaving the material is termed as the 'Critical Mass.'

Table 3.1 Energy released in a typical nuclear fission reaction

Energy component	Number per fission	Total energy (MeV)
Kinetic energy of fission fragments	2	170
Kinetic energy of prompt neutrons	2.5	5
Binding energy from capture of prompt neutrons	2.5	12
Prompt gamma rays	8	8
Total energy		195

3.2.4 Different Kinds of Fission Bombs

During the World War II, the US government started their ambitious project to develop nuclear weapons popularly known as '*Manhattan Project*' with the support of the United Kingdom and Canada. Little Boy and Fat Man are such kind of nuclear weapons, later dropped on Hiroshima and Nagasaki in August 1945 and are designed from their research and developmental project.

Complete separate methods of construction, designing and elements were applied in the making of the nuclear weapons *Little Boy* and *Fat Man*. Fission chain reaction involving isotope U-235 has detonated the Little Boy, whereas isotope Pu-239 was used in the case of *Fat Man* (Fig. 3.4).

Little Boy: It was the code name for the nuclear weapon dropped on city of Hiroshima, Japan by the United States during the World War II.

Little boy was a gun-type, fission and explosion-based nuclear weapon with a much simpler design compared to that of the Fat Boy. It derived its explosion power from Uranium-235 isotope. U-235 enrichment and extraction process have become a challenge to the Manhattan scientists as the U-235 exists in a very minute quantity (0.7 %) in the naturally found U-239. Due to the strong chemical similarities between these two isotopes, standard methods of separation cannot be used and thus made it a major task for the scientists to invent the most efficient way of separation and purification of U-235 from U-239, which exists naturally in abundance. Sufficient quantity of U-235 is required to obtain critical mass to initiate a fission chain reaction, which leads to a nuclear explosion. General Leslie Groves consulted with lead scientists of the project and agreed to investigate other methods of separation simultaneously as they were not sure of working on the new methods that could later prove insufficient to produce enough U-235. Electromagnetic separation, liquid thermal diffusion, and centrifuge and gaseous diffusion are the four separate methods investigated.

Little Boy was constructed after enough collection of isotope U-235 mainly through electromagnetic separation plant at Oakridge, Tennessee. Its detonation was based on canon mechanism that fired an amount of U-235 at another to combine the two masses which in turn would initiate a fission chain reaction. To avoid inefficient detonation, the U-235 should be rapidly assembled and two masses

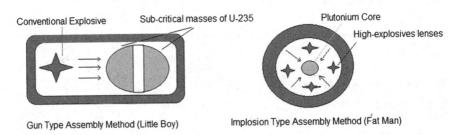

Gun Type Assembly Method (Little Boy)   Implosion Type Assembly Method (Fat Man)

Fig. 3.4 Schematic diagram of atomic bomb construction methods

of U-235 should be combined with another, fast enough to avoid the spontaneous fission of atoms. This would cause the bomb to fizzle and results in a failed bomb detonation. Of course, the gun barrel design unquestionably helped in firing a subcritical mass of U-235 atom into another down the barrel to form a super critical mass thus initiating a chain reaction.

Fat Man: It was a wide and round shaped, complex designed, and implosion-based nuclear weapon dropped on city of Nagasaki, Japan, by the United States during World War II.

It derived its implosion energy from a solid core of Plutonium-239 isotope. The form of Plutonium collected from the nuclear reactors at Hanford, WA has the traces of P-240 and hence could not use the same gun-type mechanism which allowed the Little Boy to explode efficiently. The initial Plutonium extracted from Ernest O. Lawrence's Berkeley Labs was more purified than the Plutonium collected from the Hanford labs. The main concern was before the gun-type mechanism could bring the two P-240 atoms together; they might undergo spontaneous fission due to their higher fission rate. This would lower the energy involved in the actual detonation of bomb. Therefore, a new design for the Plutonium bomb was constructed by physicist Seth Neddermeyer at Los Alamos. The Plutonium core was designed in a solid spherical shape, which facilitated the compression of subcritical Plutonium to super critical. The Plutonium core was surrounded by conventional explosives arranged in inner and outer shells to quickly squeeze and consolidate it, thus increasing the density and pressure of the substance allowing the subcritical Plutonium to reach its critical mass. Neutrons fired in this process would help in sustaining the chain reaction. When the shells were detonated, a shock wave was released that compressed the inner subcritical Plutonium core, rapidly increasing the density and making it super critical, leading to a chain reaction and finally an implosion.

3.3   Hydrogen Bomb [2]

Hydrogen bomb or *H-bomb* is a weapon, which derives its explosive energy primarily from the fission and secondarily from fusion reaction of Hydrogen isotopes. The bomb has the most devastating capability releasing tremendous amount of heat and shock once detonated. Generally, isotopes Uranium and Plutonium are used in atomic bombs, where the atoms split into smaller elements and their final weight together is less than the original atom and the difference in the mass is released as energy. Whereas, the hydrogen bomb operates on the principle of nuclear fusion in which smaller and lighter elements (Deuterium and Tritium) combine to form heavier elements (Lithium and Helium) and again the mass of the end product is less than the mass of all the lighter particles together and the mass defect is converted into energy. The hydrogen bomb requires extremely high temperature to initiate the fusion reaction and hence it is also known as *'Thermonuclear bomb.'*

As it requires high temperatures in the order of millions of degrees centigrade, the hydrogen bomb is much complex and sophisticated in design and complicated to make. Initially, a huge amount of energy is released by a nuclear fission reaction, which in turn triggers the fusion reaction. Hence, hydrogen bomb, basically, is a fusion bomb designed with a fission device to be triggered primarily. H-bombs can be made in smaller sizes and can be easily mounted into the missiles making them more convenient to design unlike the atomic bombs (Fig. 3.5).

The basic structure of the H-bomb includes a thick layer of lithium deuteride (a heavy isotope of hydrogen with a mass number 2, i.e., ^2_1H and a compound of lithium and deuterium) surrounding an atomic bomb at the center. This whole assemble is further surrounded by a thick layer made of fissionable materials (generally $^{238}_{92}U$, a radioactive isotope of Uranium-238), which binds the contents together and helps in a larger explosion. At first, the fission of lithium into helium and tritium (a heaviest hydrogen isotope with a mass number 3) is initiated by the neutrons produced from the atomic explosion. Subsequently, a fusion occurs between deuterium with tritium and tritium with tritium to produce Helium-3 protons and some energy. High temperature of 50,000,000 and 400,000,000 °C, respectively, are supplied by atomic explosion. Further, fission in the core and in the U-238 tamper is carried out by the enough neutrons produced during the fusion and the process continues. Various stages of fusion chain reactions are exoergic and thus yield energy.

The chain reaction steps are as follows:

Proton–Proton Chain: Stage 1
In the first stage of Proton–Proton chain is the fusion of two Protons to produce Deuterium (^2_1H), a Positron (e^+) and a Neutrino (v). Since the Neutrino travels close to the speed of light and interacts weakly with the other forms of matter, so they escape from the core instantly.

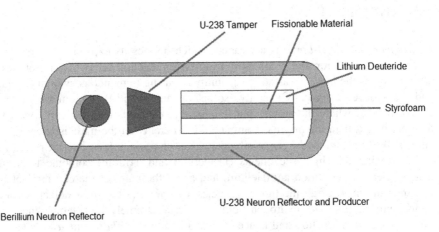

Fig. 3.5 Schematic diagram of a hydrogen bomb

$$\mathrm{_1^1H} + \mathrm{_1^1H^-} \rightarrow \mathrm{_1^2H} + e^+ + v; \quad Q = 1.44 \text{ MeV}$$

The Q—value assumes annihilation of the positron by an electron.

Proton–Deuterium Chain: Stage 2

The Deuterium obtained in the first stage could react with the second Deuterium nuclei; however, because of abundant hydrogen present the D/H ratio held low values $\sim 10^{-18}$. Thus, the next step is,

$$\mathrm{_1^1H} + \mathrm{_1^2D^-} \rightarrow \mathrm{_2^3He} + \gamma; \quad Q = 5.49 \text{ MeV}$$

Here, 'γ' indicates some energy yield carried out by gamma rays.

Helium-3–Helium-3 Chain: Stage 3

Due to the very small reaction rate with proton, helium-3 reacts with itself and burns predominantly at equilibrium. The very low concentration of deuterium makes the burning of helium with it negligible. Hence, helium-3 isotope burns with itself resulting in the formation of ordinary helium and hydrogen via the last step in the chain.

$$\mathrm{_2^3He} + \mathrm{_2^3He^-} \rightarrow \mathrm{_2^4He} + 2\mathrm{_1^1H}; \quad Q = 12.86 \text{ MeV}$$

Summarily, it is evident that hydrogen bomb is basically thermonuclear and a three-stage weapon. Fission, fusion, and again fission are the three main stages in its mechanism. A small, Plutonium-based atomic bomb (similar to the Fat Man dropped on Nagasaki) initiates the first stage, also known as a 'trigger.' Plutonium atoms undergo fission reaction and releases energy at this stage. The heaviest hydrogen isotope Tritium is also sometimes added to the core of Plutonium in order to boost the fission stage with an extra fusion energy heating the central column material to about 100 million degree centigrade. The second stage of explosion starts because of the fusion of Deuterium and Tritium. When the neutrons are released from the first stage of fission, the heavier isotopes of hydrogen become spontaneously available. They bombard with 'Lithium Deuteride,' the central column surrounding material and immense amount of energy is released. Indeed, the second stage is the thermonuclear part of the bomb. In the last stage, the incredible burst of extremely powerful neutrons made available from the second stage, split the atoms of U-238 (also termed as depleted Uranium) by fission. This is not possible at lower energy levels. Most of the radioactive radiations are released during the third stage, and this stage, in fact, doubles the explosion energy of the weapon.

Although, the manufacturing of hydrogen bomb is very cumbersome task but its destruction ability is tremendous. In the explosion of an atomic bomb, the energy released is equivalent to thousands of tons of TNT (or kilotons). However, there is as such no practical limit on the power of a hydrogen bomb. The H-bomb can be made even more powerful weapon (with the capability of releasing energy in the order of megatons) by the addition of isotopes Deuterium and Tritium at the second stage.

It should be recalled here that the H-bombs were never used till date by any nation in any war and the bombs dropped on the Japanese cities Hiroshima and Nagasaki were considered as atomic bombs. However, many countries have tested these bombs in the past and some of them are still maintaining in their arsenal. The United States successfully tested their first H-bomb in 1952, followed by Russia (then the USSR) in 1953. Later on, the thermonuclear bombs were also tested by France, United Kingdom, and China. These five nations fall under the category of Nuclear Sub-Nations and admitted to the possession and testing of these nuclear weapons, having the capability of producing such bombs and maintaining their invoice. Several other nations like Israel, India, North Korea, and Pakistan claim to either have the capability to produce them or tested these thermonuclear devices but officially state that they do not maintain the hoard of such weapons. Recently, on January 6, 2016, North Korea has claimed to have tested a miniaturized hydrogen bomb with approximately 6.0 kilotons of explosion yield and a detected quake of magnitude of 4.8.

3.4   Neutron Bombs [3]

Whenever a nuclear weapon explodes, it releases blast, heat, and ionizing radiations. The ionizing radiations have two parts: the 'prompt' radiations which include gamma rays and neutrons and the remaining portion are distributed as particles of radioisotopes which emit gamma rays, alpha and beta particles. A neutron bomb is a sophisticated radiation weapon with a low-yield of thermonuclear explosives (about 1 kiloton). It exposes the targeted area with high levels of ionizing radiations from the high energy neutrons and gamma rays.

Although, the explosion of a neutron bomb creates blast and a heat of intensity similar to that of a fission bomb but its threat to life from ionizing radiation would be at a somewhat greater scale. In the quest of cleaner weapons in terms of less radiation emittance and lesser collateral damage to the buildings, non-combatants and the environment, and the concept of neutron bomb has evolved.

Samuel Cohen invented the neutron bomb. In this fission–fusion weapon, a critical mass of Uranium or Plutonium held in place during fission by chemical explosives produces enormous amount of heat and hence maintains the optimum temperature required to initiate fusion of tritium and deuterium. The fusion reaction releases large amount of prompt radiations as neutrons. These short lived radiations in the form of neutrons are extremely damaging to the living cells and have the capability of penetrating armours and several meters of depth into the earth. In battlefield, the neutron bombs might be highly effective to target infantry formations and military tanks because of their short range destructive capacity and the lack of long range effecting capability. Also, they might not endanger the population areas and nearby cities. The neutron bombs could be possibly delivered or launched by small aircrafts and short range missiles.

References

1. Smyth HD (1945) Atomic energy for military purposes. Princeton University Press, Princeton
2. Rhodes R (1995) Dark Sun: the making of the hydrogen bomb. Simon & Schuster
3. McCally M (1991) The neutron bomb. PSR Q 1(1):4–13

Chapter 4
Missiles

Abstract Missile is the most vital component of modern day warfare. A missile is similar to a rocket which is propelled and guided to hit a specific target over varying range. They can be broadly classified depending upon their *Type, Launch Mode, Range, Propulsion System, Warhead Used,* and *Guidance System.*

Keywords Cruise missile · Ballistic missile · Launch mode · Missile aerodynamics · Missile propulsion · Missile control and guidance · Airframe · Warhead · Fuze

A Missile works on the basic principle of aiming and hitting a target with any object. A bird can be targeted by a stone and in this case the stone is considered a missile. The bird may escape the missile by changing its direction either to right or left, top or bottom with respective to the path or trajectory of the missile. This is technically known as intercepting. Here, the missile (stone) has failed to hit the target (bird) and is considered ineffective. In this case if the stone too is equipped with some intelligence and guidance to hit the bird accurately and overcome the errors, then the missile becomes a guided missile. This allows the missile to quickly respond and act in respect to the bird's evasive actions thus making it more efficient. The guided missile if used as a weapon carries armament. The armament system is a triad of missile components; warhead, fuze, and safety and arming mechanism. The modern armament triad has evolved from the necessity of three things. First, for increasing the destructive effect beyond that is available from the kinetic energy of the missile. Second, for actuating the additional destructive agent at a time such that it provides maximum damage. Lastly, to insure that destructive agent will not harm friendly personnel.

© The Author(s) 2017
M. Kaushik and P.R. Hanmaiahgari, *Essentials of Aircraft Armaments*,
SpringerBriefs in Applied Sciences and Technology,
DOI 10.1007/978-981-10-2377-4_4

4.1 Classification of Missiles [1]

4.1.1 Type

Cruise Missile: A cruise missile is a guided, unmanned (with a remote-less or preprogrammed control), single-use weapon which is designed for use against the terrestrial targets. It is a self-propelled vehicle (till the time of impact) and has the capability of traveling at supersonic Mach numbers at an extremely low-altitude trajectory with most of its flight path in the atmosphere, using a jet engine technology. Its primary mission is to place a warhead or a payload onto the target with high accuracy at longer distances. Depending upon the mode of their launch mode (air, water, submarine, etc.), speed, range, and size the missiles are classified as follows:

Subsonic cruise missile travels with a speed of around Mach 0.8 which is lesser than that of the sound. Although it takes much time to reach the target, it has an efficient engine that can dawdle over the targets (also known as 'Terrain Hugging,' which helps it to avoid the radar detection of the enemy) and can be also be self-destructed if the decision to carry out the strike is changed, thereby minimizing the possibilities of civilian casualties and collateral damages. The advantages of a subsonic missile are its capability to carry heavy payload, ability to avoid detections and its longer range. The *Harpoon* (the USA), *Exocet* (France), and *American Tomahawk* are some of the examples of subsonic missiles.

Supersonic cruise missile travels with a speed higher than that of the sound, approximately about a kilometer per second (i.e., about Mach 2 or 3). It can be launched from a wide spectrum of platforms like submarines, mobile autonomous launchers, different aircrafts, etc., due to its super modular design. When launched, a high kinetic energy ensuring a prodigious lethal effect is provided to the missile by the combination of its supersonic speed and the mass of the warhead. Also, its high speed makes it less vulnerable to the enemy defense systems and leaves a less time for the enemy camps to detect it. At present, *BrahMos-I*, a joint venture between India and Russia, is the only versatile supersonic missile system in service.

Hypersonic cruise missile travels with a speed five times faster than of the sound, i.e., about 5 mach. Although many countries are developing them, these missiles are mostly considered experimental and have not been in service so far. *BrahMos-II* is one such missile yet to be tested.

Ballistic Missile: A ballistic missile is a tube-shaped projectile, which derives its initial push from the rocket's engine and later on depends only on the gravity of the flight path. It takes a ballistic trajectory regardless of the delivery mode and the launch vehicle. After launch, a ballistic missile arches up at one point and ends at another point going up and down but never fly in the atmosphere. They have the capability of flying beyond the atmosphere like space or in vacuum. Based on the maximum distance and range calculated from the earth's ellipsoid surface from the point of launch to the point of impact of the last element of their payload, ballistic missiles are categorized into long-range or short-range missiles. The ballistic missiles can be launched from ships or submarines, aircrafts, and from land-based trucks. They are designed to carry different types of warheads such as nuclear, chemical weapons, or explosive warheads. Also, they have the capability of carrying multiple warheads in a single missile which or programmed to hit multiple targets if necessary. *Shourya* (anti-ship), *Prithvi-I, Prithvi-II, Agni-I, Agni-II,* and *Dhanush* are few ballistic missiles currently in operation.

4.1.2   Launch Mode

Surface-to-Surface Missiles (SSM): These missiles are launched from some point on the surface of the earth to another point on the surface of the earth. They could also be launched from a ship. Missiles employed for long-range targets are also known as '*strategic missiles.*' Short Range, Intermediate Range and InterContinental Ballistic Missiles (*SRBM, IRBM,* and *ICBM*) are some of the generic names (based on the range performance) of these missiles. Some examples of this type of missiles are: *CSS-3 ICBM* (China, Maximum range: 7000 km), *SS-18 Satan ICBM* (CIS-formerly USSR, 12,000 km), *Minuteman ICBM* (USA, 12,500 km), *Prithvi SRBM* (India, 100–250 km), *Agni IRBM* (India, 600–1000 km).

Surface-to-Air Missiles (SAM): Any guided missile launched from a point on the surface of the earth to destroy a target in the air qualifies for this category. The launch point, however, could be either a ship or land. Here the targets are always in motion and quite often have considerable maneuvering capability. The guidance system must be accurate since the targets are usually small in size, move at high speeds and/or are capable of executing complicated maneuvers (e.g., fighter aircraft, helicopters, SSMs). Thus, these missiles have support equipment which continuously collects information about the current position and velocity of the target. The time available for the missile to destroy a flying target is usually small and so the guidance system must be able to take appropriate actions in a short period of time. Some examples of such missiles are: *Gremlin SA-14* (CTS, 6 km), *Manpads* (France, 4–6 km), *Stringer* (the USA, 45 km), *Akash* (India, 27–35 km), *Patriot* (the USA, 160 km).

Air-to-Air Missiles (AAM): These missiles are usually launched from an aircraft to destroy targets on the surface of the earth. The targets could be either moving (but not at very high speeds) or stationary. The launch point (aircraft) is in motion. Hence, it is possible to search and seek out targets whose positions or movements are not known beforehand. In other words, the targets for such missiles are seldom predetermined as in the case of SSMs, which means that the missile must have some means of seeking out these targets. This causes the additional problems of avoiding spurious signals from the ground. Since it is possible to come close to the target, accuracy can also be improved. However, the launch point itself moves and so the velocity and other dynamic properties of the aircraft must be taken into account in the guidance system. Some examples are: *Gabriel MK-III* (Israel, 40 km), *HARM AGM-88A* (the USA, 25 km).

Air-to-Surface Missiles (*ASM*): Here, both the launch platform (air) and the target (either on land or sea) are aircrafts. These missiles are perhaps the most difficult to design and build from a guidance point of view. Both the aircrafts are at motion in high speeds. They are also capable of high maneuverability. Targets are small and difficult to locate. The guidance system has to take into account all the factors mentioned for SAMs at the target end and those mentioned for ASMs at the launch end. In addition, the guidance system should be such that it should not prevent the aircraft launching the missile from taking evasive actions for its own survival after the missile has been launched. The major advantages of ASMs are they can be fired at the target within a standoff distance and can avoid approaching too close to the target's defense limits. The attacker can just fire and forget. Some examples are: *Super 530* (France, 25 km), *Ash AA-5* (CIS, 5–20 km), *Sidewinder AIM-9* (the USA, 5–15 km).

Anti-Tank Missile: An anti-tank missile is a guided weapon primarily designed to target and destruct heavily armored military tanks. It can be operated either by a single soldier or can be launched from a tripod vehicle mount or from ships, tanks, aircrafts, etc., by a squad or team of soldiers. *Nag* is an Indian anti-tank missile currently in operation.

4.1.3 Range

A general classification of missiles based on the maximum range achieved by them is as follows (Fig. 4.1):

- Short-Range Missiles.
- Medium-Range Missiles.
- Intermediate-Range Ballistic Missile.
- Intercontinental Ballistic Missile.

4.1.4 Propulsion

Solid Propulsion: Solid propulsion uses solid propellants (fuel) for propelling the projectiles from firearms or guns. Solid propellants are made of low explosive materials mixed with high explosive chemicals in small, diluted amounts and are burnt in a controlled way without a sudden explosion. The advantages of these propellants are that they can reach higher speeds in less time, can be easily handled in fuelled state and can be stored in simple methods. Whenever a large amount of thrust is needed, these propellants make good choice due to their simplicity. Indian based *Prahaar* missile is an example of this type.

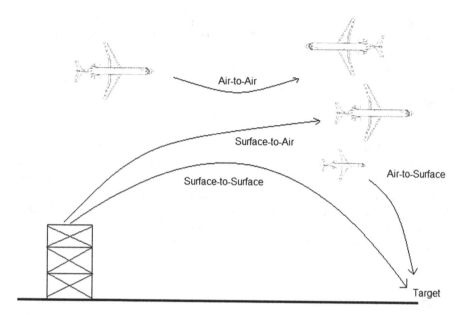

Fig. 4.1 Missile classification by method of launching

Liquid Propulsion: Liquid fuels (hydrocarbons) are used in the technology of liquid propulsion. The liquid fuels are usually filled up in the launch tanks and when they chemically react and expand in the chamber, enough thrust required for the propulsion of the projectile is produced. Under emergency conditions, the flow of the fuel can be reduced using valves which help to control the propulsion. When compared to the solid fuel it has high specific impulse but also has complex and difficult storage methods.

Hybrid Propulsion: The basic hybrid propulsion system consists of two stages, i.e., solid propulsion and liquid propulsion. The basic design has a mechanical device separating a combustion chamber (filled with solid fuel) and a pressure tank (filled with liquid fuel) where the solid propellant acts as a fuel and the liquid acts as an oxidizer. The hybrid system allows the combination of advantages and compensation of disadvantages of both the solid and liquid propulsion systems.

Ramjet: A ramjet is a type of jet engine which breathes in air and compresses this incoming air with the help of the jet's forwarded speed thus providing a simple, light propelling force for high speed flights. Unlike turbojet, it does not have any rotating components and consists of an air inlet, a combustor with a flame holder, a fuel injector, and a nozzle. The injected air sucked in through the inlet, when ignited, provides a positive push to the jet. The velocity of the exhaust air at the exit of the inlet is accelerated due to the compression and combustion of this injected fuel. At this point, the jet begins to produce thrust which pushes it forward. Ramjets always require an assisted takeoff system as they cannot push a jet engine from zero to supersonic speeds. They work efficiently at supersonic speeds.

Scramjet: Scramjet is a modification of Supersonic Combustion ramjet, (it normally uses hydrogen as fuel). The basic difference between scramjet and ramjet is that the airflow in scramjet is maintained at supersonic velocities throughout the entire path whereas the speed of the air is reduced to subsonic speeds in ramjet. Scramjet is more complex aerodynamically and much simpler mechanically than a jet engine.

Cryogenic: Cryogenics are a form of liquefied gaseous propellants stored at extremely low temperatures to maintain their liquid state. The most common examples are liquid hydrogen (acts a fuel) and liquid oxygen (acts as an oxidizer). For the gas to escape from the evaporating liquids, special insulated containers with vents (gas outlets) are required for the storage of cryogenics. From these storage tanks the liquid fuels with oxidizers are pumped into an expansion chamber. Then they are injected into a combustion chamber where they are combined and ignited using a spark or flame. The burning and expansion of the fuel produces hot exhaust gases which go out through the nozzle and provide the thrust.

4.1.5   Warhead

Conventional Warhead: A conventional warhead is a combination of high energy explosives which store a significant amount of energy in their molecular bonds. When the detonation of theses explosives is triggered, it results in the release of chemical energy and the fragmentation of the metal casing. Here, the chemical energy is the main source of explosion.

Strategic Warhead: These types of warheads are used in atomic and thermonuclear weapons which are capable of mass destruction and can be accurately delivered from one continent to other. These warheads are combination of radioactive materials which produce immense nuclear energy when detonated.

4.1.6   Guidance Systems

Wire Guidance: In this system, wires are used for passing the command signals once the missile is launched. The wires are dispensed between the missile and the guidance system located at the launch site. This system is less vulnerable to electronic measures and is similar to that of a radio command.

Command Guidance: In this system, the tracking of projectile is accomplished either by television images or radar communications relayed from the missile or by the signals transmitted by the radar or optical instruments from the launch platform. The projectile can also be tracked by transmitting commands by laser impulses, radio and radar signals, or along the thin wires or optical fibers.

Terrain Comparison Guidance (TERCOM): This type of navigation used invariably by cruise missiles involves the comparison of the terrain map (an outline map of the land and its projections and structures) to the current terrain over which the missile is flying using some sensitive altimeters which can calculate the current altitude of the missile. Thus, the missile is put on the required path by matching these altitudes.

Terrestrial Guidance: In this type of navigation, the preprogrammed angles expected on the missile's intended trajectory are compared and measured with the star angles. Whenever a variation in the trajectory of the missile is required, the guidance system directs the control system and the missile is put on the required path.

Inertial Guidance: This system is the earliest guidance system, totally pre-programmed prior to launch and does not require any support from external systems. Accelerations along the three mutually perpendicular axes are measured by three accelerometers mounted on a platform space, stabilized by gyroscopes. When these accelerations are combined twice, the first combination gives the velocity and the second one gives the position. Then the control system is directed by the guidance system to maintain the preprogrammed trajectory, which allows the missile to follow a path from which it will not deviate. Surface-to-surface and cruise missiles use this guidance system.

Beam Rider Guidance: In this system, a beam of radar energy is transmitted toward the target by an external ground or ship-based radar station. The target is tracked by this surface radar and as the target moves across the sky, the radar transmits a beam that adjusts its angle accordingly with the target and helps in the guidance of the missile.

Laser Guidance: In this system, the target is focused on by a laser beam which reflects off it and gets scattered. The laser sensor or seeker provided within the nose of the missile detects these reflections and gives direction to the guidance system. When launched, the missile works along the reflected light of the laser beam and the guidance system steers the missile accordingly toward the source of this reflections, i.e., finally toward the target.

RF and GPS Reference: In a radio frequency (RF) reference, the target is located by the missile using radio-frequent waves. Global Positioning System (GPS) allows finding out the accurate location of the target the satellites. The information collected by the missile from the GPS during its course of flight is sent as commands to the control surfaces and the trajectory of the missile is changed accordingly with the position of the target. RF and GPS are the latest missile guidance technologies used.

4.2 Missile Structure [2]

Air-to-air and surface-to-air missiles come under the category of tactical missiles. They are made of various subassemblies such as airframe, flight control system, guidance and targeting system, engine, warhead, fuze, propulsion, data link and radome.

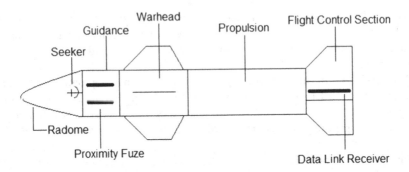

Fig. 4.2 Missile airframe

The missile components are carried by the framework known as airframe. The guidance and fuze sections are located in the forward end of the airframe. The radome covers the guidance-section seeker head to protect it from aerodynamic forces. Warhead section is located in front of the propulsion section but it is behind the guidance section. Flight control section is located wherever control surfaces are positioned. In guided missiles, the antenna and receiver are located in rear end of the airframe to have data link from the guiding unit located at the ground (Fig. 4.2).

4.2.1 Airframe

The airframes which carry the missile components are classified based on their source of lift and controls, i.e., the location of control surfaces such as wings, tails, and canards.

4.2.2 Flight Control

The flight control system uses pitch, yaw, and roll autopilots which control the airframe's motions to provide stable, controlled, and responsive missile. These autopilots are automatic feedback control systems. The pitch and yaw autopilots are also known as **lateral autopilots**.

4.2.3 Guidance

The guidance system provides steering commands to lateral autopilots that would allow the missile to reach a successful intercept of the target. The guidance system performs four major functions in order to make a decision namely seeker stabilization, target acquisition, tracking and steering signal generation.

The sensor which receives the signal which returned from the target is located in a gimballing type of system which is not affected by the missile body motions. This decoupling is achieved through a seeker head stabilization loop. Sensor motion is sensed with a rate gyroscope attached to the sensor platform. The signal from the rate gyroscope is used to generate the feedback which compensates for the body motions. The output of sensor is fed to the tracking system which keeps the sensor on tracking the target and generates steering signals. These steering commands generated by the guidance system are fed to lateral autopilots to control the airframe motions.

4.2.4 The Fuze

Every warhead must have a fuze. The fuze is that component (device) of the armament which recognizes the optimum time for detonation and initiates the detonation of the warhead at that particular time. The nature of the target and the attack geometry determines the optimum time of detonation. Fuze utilizes various means to develop information pertinent to the determination of the time of burst. These may be detectable energies generated or influenced by the target, or the calculation of the missile position by the measurement of the acceleration or time. The fuze may be physically or functionally independent of the missile system or may be partially or wholly integrated with the guidance system. Different kinds of warheads and missiles use numerous kinds of fuzes which work on different principles suitable for their environment of operation and their launch mode. Based on the activation mechanism, there are different types such as proximity, time, impact, combination, barometric and remote detonators.

Impact Fuze: Impact fuze works on the principle of detonating instantaneously or on a slight delayed detonation when the velocity of the projectile's forward motion decreases rapidly on a physical contact or by striking the target (any object or ground surface). When an armament with certain relative velocity hits another solid object, it results in a high deceleration or an inertia force, on impact. These in turn develops an electric pulse or spark which ignites the warhead. During the normal transportation and handling operations, the missile may be subjected to some impacts and the energy or force values produced in this case are much less and are not sufficient enough to trigger a detonation. The armaments are either equipped with shell-base (base detonating) or shell-nose (point detonating) fuzes. All the anti-tank missiles are provided with impact fuzes and some anti-aircraft and anti-ship missiles use them partially with proximity fuzes.

Time Fuze: These types of fuzes are time-sensitive and the detonation depends on the time period set during the launch using some chemical, mechanical, or pyrotechnic timers. Time fuze allows a missile to be either self-destructive or facilitates it to postpone the detonation by few seconds, hours, or days after it is deployed.

Proximity Fuze: These are distance-sensitive fuzes and do not require any physical contact for activation (essentially used when the possibility of impact becomes complicated due to some unavoidable errors in the missile guidance system). Prior to the missile launch, certain distance either above or below, left or right to the target is preset and the missile is expected to be in the proximity within this limit. These fuzes are basically two types. In an active fuze system, radiation is either transmitted by a very low range-power radar system or an active laser-radar system when the target is at a small distance away. Here, the fuze detonates when it receives certain strength of reflected signal. On the other hand, infra-red-based fuzes are used in passive fuze system.

Combination Fuze: These include a combination or a set of different fuzes in a parallel or a serial arrangement. This technology facilitates the munitions to explode at a certain time, distance, and when the required conditions are met.

Remote Detonators remotely control the detonation device by passing the commands through wires and radio frequency waves.

Barometric Fuzes use a certain, preset altitudes (above the sea-level) to initiate the warhead detonation. Usually, radar altimeters, infra-red altimeters, or barometric altimeters are used in sensing the altitudes.

4.2.5 Warhead

A warhead is a device made of incendiary toxic materials and explosives used to deliver destructive force on to the enemy targets which may include military bases, missile launching sites, warships, tanks, etc. These are used in military conflicts and are delivered by a rocket, torpedo, or a missile. It primarily consists of three functional units called the shell, explosive fill, and warhead case (detonator).

The primary objective of a warhead is to destroy the target by

- Converting the chemical or nuclear stored energy into destructive force.
- Producing submissiles of very high velocities.
- Submissiles which themselves have armament.
- Releasing chemical agents which corrode or damage material and poison the living organisms.
- Releasing biological or radiological agents which have destructive effects on living organisms and other materials.

The type and size of the warhead that must be employed in a given case depends on the target and the missile characteristics. Every guided missile used as a weapon, regardless of its guidance accuracy, carries a warhead. There are specifically designed warheads for different roles:

Shaped Charge: The explosive charge effect is focussed on to a hemispherical, conical, or other shaped metal hollow liner (followed by explosive on the convex side) to project a jet hyper velocity of the metal and to perforate military tanks and heavy armours (hydrodynamic penetration).

Fragmentation: Warheads made of explosive metals, when detonated scatters and projects these fragments with high velocity and pressure causing destruction and injuries to the army personnel.

Blast-cum-earth Shock: In this case, built-up structures are damaged by a shock wave produced at the time of detonation.

Incendiary: They are used to target ammunition dumps, ignite, and spread fire.

The Safety and Arming Mechanism is a transfer device between the fuze and the warhead. It provides a detonation path between the fuze and the warhead which is completed only when the S&A has determined that the detonation of the warhead cannot cause harm to their friendly personnel. If a target encounter has not already caused the fuze to detonate the warhead, the S&A sometimes additionally serve for missile self-destruction, by providing a detonation signal to the warhead or other destructive charge at some predetermined time. In a properly designed warhead and S&A combination and for an unarmed S&A state, the probability of an undesired detonation through any combination of conceivable environmental effects is very low. The unarmed condition is therefore called the "*safe condition.*" A "*premature condition*" is the warhead's blast or detonation prior to the time the S&A is designed to arm. An "*early condition*" denotes the detonation of the warhead in the period between the S&A arming and the time of the proper operation of the fuze.

4.3   Missile Aerodynamics [3]

One of the principle differences between missiles and airplanes is that the former are usually expendable and consequently are usually uninhabited. For this reason increased range of speed, altitude, and maneuvering accelerations have been opened up to missile designers and these increased ranges have brought with them new aerodynamic problems. For instance, the higher allowable altitudes and maneuvering accelerations permit operation in the nonlinear range of high angles of attack (Fig. 4.3). A missile must be ground-launched or air-launched and in consequence can undergo large longitudinal accelerations, which can also utilize very high wing loadings and can dispense with landing gear. In the absence of a pilot, the missile can sometimes be permitted to roll and thereby introduce new dynamic stability phenomena. The problem of guiding a missile without a pilot introduces considerable complexity in the missile guidance system. The combination of an automatic guidance system and the airframe acting together introduces problems in stability and control not previously encountered. Many missiles tend to be slender and may be utilized more than the usual two-wing panels. These trends have brought about the importance of slender-body theory and cruciform aerodynamics for missiles.

Some of the major points to be considered in the design optimization of the missiles:

- Reduction in both the manufacturing cost and development time of the missile by choosing a simple external configuration.
- Simplification of guidance and control systems, minimization of servo-power requirements using efficient aerodynamic control surfaces.
- Application of technology to improve speed, range, and other performance-related characteristics of the missile that can achieve the mission.
- The quality of the airframe should be adequate enough to be stable, maneuverable, and to the standpoint of dynamic responses.
- The design should comprise a power plant which is highly consistent, reliable, simple, methodical, and efficient.
- Should include airframe designs and constructions which are productive, cost-effective, and light in weight.
- Guidance and control systems should be accurate and logical enough to accomplish the desired mission.
- Authenticity and reliability of the discrete components and also the total weapon system.
- Systematic packaging of the different major components to facilitate a well-organized check-out and replacement.
- Complexity degree in the manufacturing and complications in the delivery methods of missile to fulfill its mission.

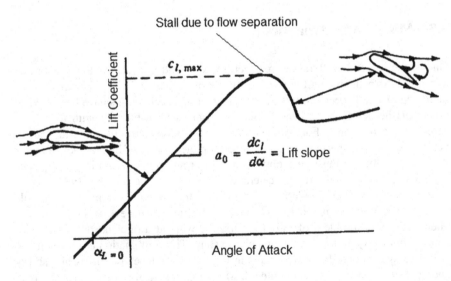

Fig. 4.3 A typical lift coefficient variation for a slender body

4.4   Missile Propulsion [4]

Propulsion is the process of providing required thrust or force to the missile to accelerate and sustain its speed if necessary, to reach the intended target. Missile propulsion system basically works on the principles of the Newton's laws of motion which are as follows:

First Law: An object continues in its state of rest or in uniform motion in a straight line unless acted upon by an external force.

Second Law: The rate of change of momentum is proportional to the impressed force and takes place in the direction of the force.

Third Law: Action and reaction are equal and opposite. That is, if a body exerts a force on another body, the other body too exerts a force on the first body of the same magnitude but in the opposite direction.

A rocket engine is a kind of a reaction engine which generates thrust by ejecting hot exhaust gaseous fluids called 'propellants' from its nozzle which assists in the propulsion of the missile. The solid or liquid propellants (energetic fuels) with the combination of some oxidizers undergo chemical reactions in the combustion chamber producing hot gases. The nozzle releases these exhausted gases with supersonic speeds and as a reaction these in turn push the engine in opposite direction, implementing the third law of Newton. As per the law of momentum (Newton's second law), a force is developed due to the change in the rate of momentum in the body of missile including the rocket motor casing, nozzle, and the entire system. This force leads to the propulsion of the missile. Once propelled into the air, the missile is vulnerable to other forces like gravitational force acting downwards toward the center of earth and may suffer an aerodynamic drag (resistance to its forward motion due to the air). To overcome this opposite or negative forces, law of inertia (Newton's first law) is applied and the missile is imparted with some compensative forces to continue its forward movement.

4.4.1   Uses of Propulsion Systems

All the missiles, warships, space crafts, satellite-launching vehicles are launched by a rocket engine which works on the concept of the propulsion systems. The lethal capacity of a missile is determined by the technology of its warhead guidance accuracy. Every country closely guards this technology and maintains this information as classified. However, in some civil applications and operations such as weather forecasting, mineral surveys, meteorology, mapping and satellite communications the rockets with satellite payloads are used.

4.5 Missile Control and Guidance [5]

Guided weapon is a generic term which applies to any of the ballistic systems when they are deliberately perturbed from their ballistic path after launch, in order to increase the probability of hitting the target. There are three fundamental problems in guided weapons:

- Determining where the target is or will be.
- Determining where the weapon is.
- Correcting the weapon's location to coincide with the target's location at the time of closest encounter.

The former two problems come under guidance and the latter under control. Control is usually accomplished by aerodynamic-control-surface deflection.

There are five fundamental concepts of guidance namely,

- Inertial guidance
- Command guidance
- Active guidance
- Semi-active guidance
- Passive guidance.

In command guidance, the weapon system operator or onboard sensors observe the relative location of the weapon and the target and then direct the trajectory corrections. In an active guidance system, the electromagnetic emissions, for example, radar, microwave, or laser are transmitted from the weapon to the target and the return energy reflections are measured to determine the range and the angle with respect to the target. Semi-active guidance resembles active guidance except that the illumination of the target is provided by a designator not located on the weapon. Passive guidance uses the natural emissions radiating from the targets to uniquely acquire a target and subsequently guide the weapon to the target. The ultimate objective of a weapon system is to be completely autonomous. An example of an autonomous system is combining the inertial and passive or active terminal guidance with the appropriate algorithms to acquire the target after launch without the operator's intervention. In this case, the weapon is launched into an area where targets are known to exist and upon reaching the area, the weapon searches and finds the target homes on its own and destroys them. The trend toward weapons which can autonomously acquire the targets allows the building of weapons with substantial standoff capability. This increases the survivability of the launch platform, improves accuracy, increases proficiency, and reduces the logistical burden. There are two classes of smart weapons. The first class consists of those guided weapons that possess some form of terminal guidance and home in on the target. Weapon systems in this class include *laser-guided bombs*. The second class of smart weapons includes those that autonomously acquire the target after launch and are usually termed lock-on-after-launch, fire-and-forget, or brilliant weapons.

As already mentioned, the direction of the missile in which it should move to reach the intended target is decided by the aspect of *missile guidance system*. During the entire flight of the missile, the guidance system takes decisions at every regular short interval. Hence, the more accurate the guidance system, the more effective is the missile. Various types of guidance systems are as follows:

- Command guidance
- Homing or seeker guidance
- Beam rider guidance
- Inertial guidance and
- Stellar guidance.

4.5.1 Control Force Generation

Methods like thrust vector control, aerodynamic control, vernier rockets, and reaction control systems are used for producing adequate force required to generate a turn in the missile.

4.5.2 Aerodynamic Control

This method works on the simple principle that if the speed of air on a surface is increased then the pressure exerted on the surface by the air decreases, keeping the total energy a constant. This method works for missiles traveling in the atmosphere with certain minimum speed and here the body of the missile is provided with flat aerodynamic surfaces called *control surfaces*. A local differential force is produced when these surfaces are diverted with respect to the missile's body. This force generates moment acting on the body resulting in its rotation about a particular axis. Canard control (nose end), moving wind control (middle) and tail control are the different types of control surfaces termed, based on their location along the longitudinal axis of the missile. Each of it has its own specific applications and advantages and the control force generated is a function of the size of control surface, dynamic pressure, shape, and angle of deflection. The dynamic pressure, in turn, is a function of the density of air at the missile's altitude and its velocity. The center of gravity of the missile, location of the pressure of aerodynamic forces acting on the body and the control force, altogether determine the turning moment of the missile.

4.5.3 Thrust Vector Control

In this method, the direction of the thrust force vector is diverted and the control force and moment are generated. The thrust is manipulated by insertion of inner and outer blades at the jet's exit, nozzle rotation (a flexible nozzle is used in solid rocket motors), or by gimballing the engine (in liquid propellant engines). This system cannot be applied when the engine stops burning and can be used when the missile's velocity is not adequate during its launch and when it travels in space or in low density atmosphere.

4.5.4 Reaction Control System or Vernier Rockets

These are generally liquid propellant systems based on the concept of chemical propulsion and can be switched on and off when required. In this system, the main engine is additionally provided with a number of sets of independent small thrust body-fixed engines to provide control along various axes.

4.6 Ballistic Missile [6]

This missile is designed primarily in accordance with the laws of ballistics, i.e,. non-lifting gravity turn trajectory with an angle of attack near zero. It is a self-propelled guided missile for surface targets that depends primarily upon the thrust of its propelling system rather than the aerodynamic lift. Most part of the trajectory is in the outer atmospheric region. The trajectory is essentially a parabola. Ballistic Missiles consist of three phases namely *powered flight, ballistic flight,* and *re-entry phase* (Fig. 4.4).

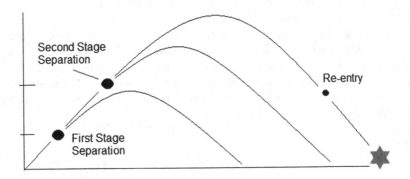

Fig. 4.4 Trajectory of a long-range missile

Table 4.1 Ballistic missile classification based on range

Missile	Range
Medium range ballistic missile (MRBM) Ex. *CSS-6, Agni-1, Hatf-4 (Shaeen-1), Scud-C*	500–2000 km
Intermediate range ballistic missile (IRBM) Ex. *CSS-5, Agni-2, Taep'o-dong-1, Gauri-2*	2000–5000 km
Intercontinental range ballistic missile (ICBM) Ex. *CSS-9, Taep'o-dong-1&2, SS-18, SS-19, Peacekeeper*	>5000 km

4.6.1 Characteristics of Long-Range Ballistic Missiles

For optimum range, multi-staging is required. Missiles are generally launched vertically from launching pad (Table 4.1).

- After few seconds of vertical rise missile is commanded into a gravity (or Zero lift) turn, tilting the velocity vector of the missile.
- After sometime the missile is commanded to fly over a constant attitude trajectory.
- Powered flight is continued in this predetermined trajectory until the required injection velocity and flight path angle are achieved.
- Once it achieves required burn out velocity, the payload is separated from the last stage and commences its free flight ballistic trajectory to the target.
- Lift is generally produced by body alone.
- Lifting surfaces of small size are sometime provided to augment stability and control.
- Most of the flight is in the upper atmosphere (no drag).
- The angle of attack during the flight is very low.
- High Mach number regions (Mach up to 16).
- Higher Slender ratio (l/d about 20) leading to flexibility of the vehicle.
- Reentry aerodynamics encountering severe aerothermal environment.

4.6.2 Need for Staging

For a given payload mass, in order to minimize the overall mass, staging is employed (or) to maximize the payload for a given overall mass.

$$\Delta V = I_{SP}g \ln\left(\frac{m_0}{m_{payload} + m_{structure}}\right)$$

$$\Delta V = I_{SP}g \ln\left(\frac{m_0}{m_{payload} + \sigma\left(m_0 - m_{payload}\right)}\right)$$

$$\text{Structural factor, } \sigma = \frac{\text{Structural mass}}{(\text{propellant mass } + \text{ structural mass})}$$

Rearranging the above equations, the ratio of overall mass to payload mass is given by,

$$\left(\frac{m_o}{m_{\text{payload}}}\right) = \frac{(1 - \sigma)}{\left(e^{-\frac{\Delta V}{I_{SP}g}} - \sigma\right)}$$

Choosing a value for the structural factor (σ), stage weight and thus the propellant weight is obtained from the above equation. For same structural factor and same specific impulse, the mass ratio,

Single Stage:

$$\left(\frac{m_o}{m_{\text{payload}}}\right) = \left(\frac{(1 - \sigma)}{e^{-\frac{\Delta V}{I_{SP}g}} - \sigma}\right)$$

Two Stages:

$$\left(\frac{m_o}{m_{\text{payload}}}\right) = \left(\frac{(1 - \sigma)}{e^{-\frac{\Delta V}{2I_{SP}g}} - \sigma}\right)^2$$

Three Stages:

$$\left(\frac{m_o}{m_{\text{payload}}}\right) = \left(\frac{(1 - \sigma)}{e^{-\frac{\Delta V}{3I_{SP}g}} - \sigma}\right)^3.$$

4.7   Problems Associated in Missile Launch

A missile launch concept is based on the principles of aerodynamics which always face a few challenging tasks. The missile and the aircraft's design compatibility deeply influence the efficiency of the missile. The launch platforms of the missile: from ground (static or mobile), from water (warships), or by high-speed aircrafts and the design of the missile are some of the challenging factors which surface frequently during the firing of the missile. In order to minimize the undesirable launching dispersion caused by these external forces, a detailed analytical and logical study of both the magnitude and source of the internal and external forces acting on the missile during its launching stage is essential. The missile compatibility of the aircraft must be considered in the case of air-launched missiles. A successful launch of a missile can be assured by incorporating modifications to the missile's design technology. Enough care should be taken during the launch that the missile's launching dispersion does not exceed the limit proposed by the guidance considerations. This can be achieved by assuring the parent aircraft is safe for launch and meets the necessary requirement. When the aircraft and missile designers work separately as two distinct groups then a state of '*frozen design*' arises at which the retro-fitting of the missile to the aircraft becomes complicated. Application of some modifications to the design also becomes difficult at this state.

Based on safety criteria for both the aircraft and missile designers, the following points should be taken into consideration:

- Under any condition of the flight, the missile should not fail in the immediate vicinity of the parent aircraft and the missile should not hit the aircraft during jettison or boost.
- The operating components of the parent aircraft such as air inlets and control surfaces and its structure should not be adversely affected by the jet blast from the missile's rocket.

Problems involved in Ground launch are categorized into

- Influence and consequences of the launching stage on the missile.
- Effects of the missile on the launch platform and surrounding areas.

The effects of the launching phase on the missile are studied from the standpoint of the missile component exercises and the guidance system. Whereas, the effects of missile on the launching site are taken into consideration to assure the safety of the crew members and civilians in the surrounding areas. These effects can be minimized by avoiding the excessive missile dispersion launch. Many factors like fin misalignments, atmospheric disturbances like tail wind, cross wind, gusts, missile tip-off from the launcher and thrust cause the dispersion.

4.7.1 Launchers and Ground Support Systems

Launchers are the most important factors in the missile operating exercises to reach a specific and intended target. Certain ground systems are essential for the launch of any missile. For example, a launch-cum-container tube resting on the human shoulder is used for firing the small missiles, whereas, military vehicles such as submarines (under water), silos (underground), warships (above water) and mobile vehicle launchers are used for launching large ballistic missiles. The design and selection of the launcher involves an application of innovative technology and engineering skills for a precise and accurate aiming of a particular target. In the case of anti-aircraft missiles, the launchers play a vital role due to the high rates of fluctuations in azimuth and elevation.

When a missile is deployed, its range can be immensely tested. *Development testing* and *user evaluation testing* are the two phases of a missile range test. The missile test ranges should be safe enough to carry out the experiments without harming the civilian areas. The flight path, size, and range of the missile influence the extent of safety zones in these locations. Some of them are located in the deserts and few of them near to the sea. The ranges should have adequate data collecting instrumentation and lab facilities to evaluate the range and flight of the missile.

In India, the launch vehicles testing ranges equipped with instrumentation and other facilities are present in *Sriharikota, Chennai,* and *Trivandrum.* A major range facility is located at *Balasore* (Odisha). Electro-optic instruments, telemetry receiving stations, meteorological facilities and tracking radars are the main instrumentation facilities provided at these testing locations. For carrying out tests of flight vehicles using telemetry command system, real-time data processing facilities are in existence.

References

1. Kapoor AK, Karthikeyan TV (1990) Guided missiles. Popular Science and Technology Series. DESIDOC. Defence Research and Development Laboratory
2. Carleone J (1993) Tactical missile warheads. Progresses in astronautics and aeronautics, vol 155
3. Nielsen JN (1960) Missile aerodynamics. Nielsen Engineering & Research, Inc., California
4. Netzer D, Jensen J (1996) Tactical missile propulsion: progress in astronautics and aeronautics. American Institute of Aeronautics and Astronautics Publishing
5. Siouris GM (2004) Missile guidance and control systems. Springer
6. Steven A (2003) Ballistic missile 1942–52, 2nd edn. Osprey Publishing

Chapter 5
Armament Materials

Abstract This chapter briefly reviews the properties of various materials required in the production of armaments. The 'gun powder' or 'black powder' is the earliest known explosive material used in gun barrels. Armaments are made up of those materials which are corrosion and chemical agent resistant. The metals with higher strength-to-weight ratio are advantageous for armament applications.

Keywords Armament material · Material strength · Corrosion resistance · Stiffness · High-temperature properties

In order to qualify for the production of armaments, a metal must have following material properties:

- High material strength
- Higher strength to weight ratio
- High-temperature properties of the material
- Stiffness or deformation characteristics
- Corrosion and chemical agent resistant
- Ease of fabrication.

The armament materials can be broadly classified on the basis of burning rate. First category includes those explosives in which decomposition of materials takes place though a flame front traveling at subsonic speeds (also known as Deflagration), are termed as 'low-explosives'. The common example of this type is 'Gun Powder'. It is the earliest known armament material used in China around 1000 AD. It is a mixture of fuel (sulfur or charcoal) and oxidizer (potassium nitrate), when ignited behind the bullets it generates enough pressure to eject them from the muzzle at high velocities. On the other hand, when the decomposition of explosive takes place through a 'shock wave' traversing the explosive material at supersonic speeds (termed as Detonation) suddenly releases abundant light, energy, and sound arc known as 'high explosives.' Trinitrotoluene (TNT) comes under this category. The atomic bomb is another example of 'high explosives' which uses fissile

© The Author(s) 2017
M. Kaushik and P.R. Hanmaiahgari, *Essentials of Aircraft Armaments*,
SpringerBriefs in Applied Sciences and Technology,
DOI 10.1007/978-981-10-2377-4_5

material as isotopes of Uranium (U^{235} and U^{238}), once detonated undergoes in a chain reaction producing enormous heat.

For the production of missiles, the metals like magnesium, titanium, aluminum, and its alloys are extensively used. The usage of new metals in the missile technology has come into existence due to the fact that the missiles travel at supersonic speeds and encounter extreme temperatures during their flight. As a result, lighter metals such as *molybdenum, beryllium, graphite compounds,* and *fiber-reinforced plastics* like the carbon–carbon variety are being used in a large scale.

Chapter 6
The United Nations

Abstract The organization of United Nations is reviewed in the present chapter. Chronological development and current status of Non-Proliferation Treaty (NPT) of nuclear arsenal is presented. The Geneva Protocol (1925), Biological Weapon Convention (1972), and Chemical Weapon Convention (1993), as adopted by the United Nations General Assembly, in order to prohibit the use of bacterial and toxic weapons against other member countries are also discussed.

Keywords United Nations · Non-proliferation treaty · Biological weapon convention · Chemical weapon convention · Nuclear-suppliers group

6.1 United Nations Organization [1]

In World War II, humanity witnessed the devastation of lives and property in a large scale. It was the first time when world has seen the most unconventional weapons in warfare with greater capabilities of killing people. Thousands of lives were lost particularly during atomic explosions at Hiroshima and Nagasaki. After the war, there was a strong urge to foster peacekeeping process and eradicate any possibility of wars in future. In the view of this, the United Nations (UN), a unique organization of independent countries came together to work for peace and prosperity. On October 24, 1945, a total of 51 countries joined hands in forming UN and by the end of 2008 its membership grew up to 192 countries. The UN is headed by an elected representative termed as 'General Secretary.'

The Charter of UN is a set of guidelines that explains the rights and duties of member countries. It was signed on June 26, 1945, in San Francisco and came into force on October 24, 1945. The Statute of the International Court of Justice is an integral part of the Charter. Presently, the Charter consist a total of 111 articles spread over 14 chapters.

The UN has four main purposes to spread peace and social progress.

- Maintaining peace throughout the world.
- Developing cordial relations between member countries.

M. Kaushik and P.R. Hanmaiahgari, *Essentials of Aircraft Armaments*,
SpringerBriefs in Applied Sciences and Technology,
DOI 10.1007/978-981-10-2377-4_6

- To work for helping poor people to live better lives; to conquer hunger, diseases, and illiteracy; to encourage respect for each other's freedom and sovereignty.
- To work as center in order to achieve the above goals.

The United Nations function through its six main organs.

- General Assembly
- Security Council
- Economic and Social Council
- Trusteeship Council
- International Court of Justice
- Secretariat.

Except the International Court of Justice, which is located at The Hague (Netherlands), all other organs are based at UN Headquarters in New York. The land on which UN headquarter sits does not belong only to the host country, i.e., USA, but it is an international zone guarded by the UN employed security officers.

The official languages used at UN are Arabic, Chinese, English, French, Russian, and Spanish. The working languages at UN Secretariat are English and French. A delegate may speak in any of the six official languages, and the speech is interpreted simultaneously into other official languages. The delegate is allowed to speak in any nonofficial language but he or she has to provide its translation in any of the six official languages.

6.2   The Role of United Nations in Peacekeeping [2]

The establishment of UN aims to foster peace and security throughout the world. It plays a pivotal role in resolving conflicts, and enforces law as described in the various charters of its constitution. The role of UN is very vast and cannot be encapsulated in a single text, and hence we will confine our discussion in understanding its importance in non-proliferation of weapons of mass destruction through various treaties.

On June 17, 1925, the first consolidated effort was made by the global community in prohibiting the use of biological weapons and toxic gases in warfare. It was termed as Geneva Protocol which came into force on February 8, 1928. Member countries pointed out that they only accept the first nonuse obligation of biological and chemical weapons against others; however, the protocol will cease to apply as soon as they are attacked by these prohibited weapons. It was a general law which prohibited the use of chemical and biological weapons but it was silent on storage and proliferation of these weapons. Later, these aspects were covered by Biological Weapon Convention (BWC) in 1972 and Chemical Weapon Convention (BWC) in 1993.

In 1970, the international community signed a treaty at UN, commonly called Non-Proliferation Treaty (NPT) of nuclear weapons. Its prime objective is to prevent the spread of nuclear technology for destructive purposes, and rather promoting its use for satisfying enormously increasing energy needs. Till date, a total of 191 member countries have signed on the NPT. North Korea has accepted the treaty in 1985 but never adhered to it, and finally withdrew in 2003. Four member states, India, Israel, Pakistan, and South Sudan have never signed on it. This treaty recognizes five member states United States of America, Russia (Formerly Soviet Union), United Kingdom, and France as nuclear weapon capable states. These countries are the permanent members of UN Security Council (UNSC) as well. In every 5 years, the NPT is reviewed in the review conferences. On June 24, 2016, at Seoul conference, India again failed in entering the Nuclear Suppliers Group (NSG) due to strong opposition led by China, and supported by Brazil, Switzerland, Mexico, and Turkey citing India's non-adherence to NPT. The NSG is a group of member states who are instrumental in preventing the nuclear proliferation through various checks on spreading the material, equipment, and technology. At present, there are a total of 48 member countries in NSG. As a precondition, only those nuclear weapons capable states, who have signed on NPT, can be a member of NSG. Though India has shown its capability of having nuclear arsenal through Pokhran-I (1974) and Pokhran-II (1998) nuclear tests but it has not yet signed on NPT.

Biological and Chemical weapons are the other major threats under weapons of mass destruction category, whose impacts are comparable or even stronger than nuclear weapons. After many rounds of negotiations, the United Nations disarmament forum proposed a written document formally known as Biological Weapon Convention (BWC), which was later adopted by general assembly in 1972. It came into existence after its ratification by 144 member states in 1975. This treaty was evolved to support the Geneva Protocol of 1925, which prohibited the production and agglomeration of bacterial and toxic weapons. BWC empowers a member state to file a complaint to UNSC in case of any suspicion that a member country is violating the norms as stipulated under this treaty. It also gives a right to UNSC to investigate the matter and to resolve it. However, because of the vested veto power in all the five member states of UNSC the investigation could not be carried out so far, despite the fact that some member states like Soviet Union, North Korea, Iran, and Libya were assumed to violate the BWC norms.

On April 28, 2004, the United Nations Security Council has adopted its resolution 1540 under Chapter VII of UN Charter, creating an obligation for the member countries to modify their legislation suitably in order to have non-proliferation of weapons of mass destruction.

During the cold war era, the proliferation of chemical weapons and its use at a large scale during Iraq–Iran War by Saddam Hussein against Kurdish people forced the international community to work effectively for its eradication and non-proliferation. Similar to Biological Weapon Convention (BWC), the Chemical Weapon Convention (CWC) was also framed by disarmament forum and later adopted by United Nations General Assembly in 1993. A total of 188 member countries acceded to abide by this convention. Unlike BWC, the rules stipulated in

CWC are more stringent and enforceable. The member states have to undergo the registration of their chemical-related activities or material possessed, and the inspection and close monitoring make sure the compliance of CWC.

References

1. The United Nations (2016) Available at: http://www.un.org/en/sections/about-un/overview/index.html. Accessed 28 June 2016
2. Morris B (1991) World armament and disarmament. Stockholm International Peace Research Institute, p 296

Appendix A

A.1 Multiple Choice Questions and Answers

1. What are the aerodynamic forces acting on a missile when it is airborne?

    (a) Lift
    (b) Drag
    (c) Both a & b
    (d) None

2. The major lift producing structure on missile is called

    (a) Fin
    (b) Canard
    (c) Wing
    (d) Body

3. Lateral acceleration is defined as

    (a) Total normal force/mass
    (b) Thrust/mass
    (c) Mass/axial force
    (d) Axial force/normal force

4. What is the inertia force of a vehicle of mass 100 kg that experience a latex of 5 g?

    (a) 20 N
    (b) 50 N
    (c) 500 N
    (d) 5000 N

5. When missile is at trim condition what happens?

    (a) Both CG and CP coincide
    (b) No unbalanced moments
    (c) Both a & b
    (d) Nothing happen

© The Author(s) 2017
M. Kaushik and P.R. Hanmaiahgari, *Essentials of Aircraft Armaments*,
SpringerBriefs in Applied Sciences and Technology,
DOI 10.1007/978-981-10-2377-4

6. Selection of a material for missile depends on

    (a) High specific strength
    (b) High specific stiffness
    (c) Both a & b
    (d) Only low cost

7. When aluminum, steel, and titanium of same size meet the requirements of design, which material will you choose?

    (a) Steel
    (b) Aluminum
    (c) Titanium
    (d) A combination of two materials

8. Why is load analysis required?

    (a) To evaluate design load
    (b) To calculate temperature
    (c) To finalize the aerodynamic configuration
    (d) To provide proper damping

9. Which is the first step of airframe design?

    (a) Load analysis
    (b) Trajectory analysis
    (c) Material selection
    (d) Thermal analysis

10. Flutter is a _____ instability

    (a) Static
    (b) Dynamic
    (c) Quasi-static
    (d) Steady state

11. BrahMos-I is a _____ missile

    (a) Subsonic
    (b) Supersonic
    (c) Hypersonic
    (d) Transonic

12. Divergence is due to interaction between

    (a) Aero and inertia
    (b) Mass and inertia
    (c) Aero and elastic
    (d) Aero, elastic and inertia

13. Vibration is due to interaction between

 (a) Elastic and inertia
 (b) Aero and inertia
 (c) Mass and inertia
 (d) Aero and elastic

14. Buffeting is _____oscillation?

 (a) Harmonic
 (b) Turbulent
 (c) Steady
 (d) Stream line

15. A spring mass system has a weight of 10 kg and spring stiffness of 1000 N/m. Its frequency is

 (a) 10 rad/s
 (b) 100 rad/s
 (c) 10,000 rad/s
 (d) 1 Hz

16. What is modal analysis?

 (a) Determination of frequency
 (b) Determination of mode shape
 (c) Both a & b
 (d) Determination of buckling load

17. Why modal analysis is required?

 (a) Positioning of sensor
 (b) Designing filters
 (c) Both a & b
 (d) Selection of materials

18. To improve the frequency of a structure, you

 (a) Increase stiffness
 (b) Decrease mass
 (c) Both a & b
 (d) Increase mass

19. The famous Tacoma suspension bridge failed due to

 (a) Divergence
 (b) Flutter
 (c) Vibration
 (d) Fatigue

20. Flutter is a

   (a) Forced vibration
   (b) Self-excited oscillation
   (c) Undamped oscillation
   (d) Random oscillation

21. Mach number of hypersonic vehicle

   (a) <1
   (b) >5
   (c) <2.5
   (d) 0

22. Which of following has got maximum strength to weight ratio?

   (a) 15-5 PH steel
   (b) Ti
   (c) Al
   (d) Composites

23. Which of following is a hypersonic vehicle program in DRDO?

   (a) NAG
   (b) LRSAM
   (c) HSTDV
   (d) None

24. Dynamic pressure is given as

   (a) $1/2\ v\rho^2$
   (b) $1/2\ \rho v^2$
   (c) $1/2\ \rho v$
   (d) None

25. Agni 5 missile range is

   (a) 2000 km
   (b) 5000 km
   (c) 3000 km
   (d) 500 km

26. Mach number is the ratio of object speed to speed of sound

   (a) True
   (b) False

27. Crank and lever operation in an automobile is a

   (a) Structure
   (b) Mechanism

(c) Both

(d) None

28. If stiffness and mass of a system is doubled, the frequency will be

(a) Doubled

(b) Half

(c) Unchanged

(d) Can not say

29. Structure is the one which has got no relative motion

(a) True

(b) False

30. For stability point recommended position of CG of a control surface

(a) Ahead or on hinge line

(b) After hinge line

(c) Anywhere

(d) None

31. What is the purpose of conducting structural testing on missile components?

(a) Validating design

(b) Prove structural integrity

(c) Validating fabrication process

(d) All the above

32. Name the sensor from the list of instruments given below?

(a) Load cell

(b) COD gage

(c) Strain gauge

(d) Pressure transducer

33. What structural test validates design and fabrication processes?

(a) Static

(b) Dynamic

(c) Qualification tests

(d) Acceptance tests

34. Classification of structural tests based on severity of load applied

(a) Qualification tests

(b) Static structural tests

(c) Acceptance tests

(d) Both a & c

35. What kind of loads are experienced by missiles during their development?

    (a) Flight loads tests
    (b) Transportation and handling loads
    (c) Internal pressure
    (d) All of a, b & c

36. What is the diameter and range of A5 missile?

    (a) 1.2 m and 4000 km
    (b) 1.2 m and 3000 km
    (c) 2 m and 3000 km
    (d) 2 m and 5000 km

37. What is the unit of strain measured during the structural tests?

    (a) Microns
    (b) No units
    (c) MPa
    (d) Both a & c

38. Name of the equipment or device which apply load on the structure

    (a) Actuator
    (b) Test rig
    (c) Load cell assembly
    (d) Fixture

39. Type of hydraulic actuator which can apply push and pull loads

    (a) Sing acting
    (b) Both a & c
    (c) Double acting
    (d) None of the above

40. Name of the transducer which can measure joint rotation constant

    (a) Strain gage
    (b) LVDT
    (c) Load cell
    (d) COD gage

41. Natural frequency of a structure depends upon

    (a) Mass and stiffness
    (b) Mass and damping
    (c) Stiffness and damping
    (d) None of the above

42. Sensing element in accelerometer is

    (a) Piezoelectric crystal
    (b) Strain gage
    (c) Piezo resistive
    (d) All above

43. What is the boundary condition of missile in flight?

    (a) Cantilever
    (b) Fixed-fixed
    (c) Free-free
    (d) Simply supported

44. What is the purpose of charge amplifier?

    (a) To convert high impedance charge signal to low impedance voltage signal
    (b) To provide matching sensitivity
    (c) To provide and low and high pass filters
    (d) All above

45. Vibrations can be measured

    (a) Displacement
    (b) Velocity
    (c) Acceleration
    (d) All of the above

46. The Young's modulus of most of the metallic materials used in aerospace applications varies with temperature. With increase in temperature it

    (a) Increases
    (b) Decreases
    (c) First increases up to critical temperature and then decreases
    (d) Remains constant

47. The part of the electromagnetic spectrum responsible for heating up of objects is

    (a) X-rays
    (b) Visible light
    (c) Ultraviolet rays
    (d) Infrared

48. The filament material used in almost all infrared lamps is

    (a) Tungsten
    (b) Titanium
    (c) Chromel
    (d) Inconel

49. Consider the following statements:

> I. Aerodynamic heating can take place in space
> II. Design of airframes at room temperature is adequate for development of missiles
> III. Infrared heating using IR lamps are commonly used for thermo-structural testing

Which of the above is/are true?

(a) Only I
(b) I and II
(c) Only III
(d) All of the above

50. The emissivity of an ideal black body is

(a) 0
(b) Less than 1
(c) Equal to 1
(d) Greater than 1

51. Total number of members in United Nations Security Council (UNSC) is?

(a) 4
(b) 5
(c) 7
(d) 11

52. The Non-Proliferation Treaty (NPT) of nuclear weapons was adopted by the UN in the year

(a) 1925
(b) 1945
(c) 1970
(d) 2004

53. The Biological Weapon Convention (BWC) was adopted by the UN in year

(a) 1970
(b) 1972
(c) 2004
(d) 2015

54. The Chemical Weapon Convention (CWC) was adopted by the UN in year

(a) 1970
(b) 1975
(c) 1993
(d) 2014

55. The total number of articles in UN constitution is

   (a) 55
   (b) 100
   (c) 110
   (d) 111

56. The Geneva Protocol came into existence in the year

   (a) 1925
   (b) 1945
   (c) 1970
   (d) 2007

57. Which of the following country is not having veto power?

   (a) China
   (b) France
   (c) Russia
   (d) Germany

58. The Nuclear Suppliers Group (NSG) has member states

   (a) 48
   (b) 111
   (c) 188
   (d) 192

59. Which of the following countries initially signed on NPT in 1985 but later
withdrew?

   (a) India
   (b) Pakistan
   (c) North Korea
   (d) South Korea

60. Which of the following is not the official language of UN?

   (a) Arabic
   (b) Chinese
   (c) English
   (d) Italian

Appendix B

B.1 Multiple Choice Questions

Answers

1	c	31	d
2	c	32	c
3	a	33	c
4	d	34	d
5	c	35	d
6	c	36	d
7	b	37	b
8	a	38	a
9	b	39	c
10	d	40	d
11	b	41	a
12	c	42	d
13	a	43	c
14	b	44	d
15	a	45	d
16	c	46	b
17	c	47	d
18	c	48	a
19	b	49	c
20	b	50	c
21	b	51	b
22	d	52	c
23	c	53	b

(continued)

© The Author(s) 2017
M. Kaushik and P.R. Hanmaiahgari, *Essentials of Aircraft Armaments*,
SpringerBriefs in Applied Sciences and Technology,
DOI 10.1007/978-981-10-2377-4

(continued)

24	b	54	c
25	d	55	d
26	a	56	a
27	b	57	d
28	c	58	a
29	a	59	c
30	a	60	d

Bibliography

1. Bruce JM (1965) Warplanes of the First World War—fighters, vol 1. MacDonald & Co., London
2. Needham J (1986) Military technology: the Gunpowder Epic. Cambridge University Press, Cambridge, pp 180–181
3. Nielsen JN (1960) Missile aerodynamics. Nielsen Engineering & Research, Inc., California
4. Shaw RL (1985) Fighter combat: tactics and maneuvering. Naval Institute Press, Maryland
5. Wahl P, Toppel D (1971) The Gatling gun. Arco Publishing
6. Smyth HD (1945) Atomic energy for military purposes. Princeton University Press, Princeton
7. Rhodes R (1995) Dark Sun: the making of the hydrogen bomb. Simon & Schuster
8. McCally M (1991) The neutron bomb. PSR Q 1(1):4–13
9. Kapoor AK, Karthikeyan TV (1990) Guided missiles. Popular Science and Technology Series. DESIDOC. Defence Research and Development Laboratory
10. Carleone J (1993) Tactical missile warheads. Progresses in Astronautics and Aeronautics, vol 155
11. Siouris GM (2004) Missile guidance and control systems. Springer
12. Megson THG (2007) Aircraft structures for engineering students, 4th edn. Butterworth-Heinemann
13. Steven A (2003) Ballistic missile 1942–52, 2nd edn. Osprey Publishing
14. Netzer D, Jensen J (1996) Tactical missile propulsion: progress in astronautics and aeronautics. American Institute of Aeronautics and Astronautics Publishing
15. Cheremisinoff NP (1997) Handbook of engineering polymeric materials. Marcel Dekker Inc., New York
16. The United Nations (2016) Available at: http://www.un.org/en/sections/about-un/overview/index.html. Accessed 28 June 2016
17. Morris B (1991) World armament and disarmament. Stockholm International Peace Research Institute, p 296

© The Author(s) 2017
M. Kaushik and P.R. Hanmaiahgari, *Essentials of Aircraft Armaments*,
SpringerBriefs in Applied Sciences and Technology,
DOI 10.1007/978-981-10-2377-4

Printed in the United States
By Bookmasters